WHY WE FIGHT

WHY WE FIGHT

MORAL CLARITY AND THE WAR ON TERRORISM

WILLIAM J. BENNETT

Since 1947
REGNERY
PUBLISHING, INC.
An Eagle Publishing Company • Washington, DC

First paperback printing 2003

Published by arrangement with The Doubleday Broadway
Publishing Group, a division of Random House, Inc.

Cataloging-in-Publication Data on file with the Library of Congress

Published in the United States by
Regnery Publishing, Inc.
An Eagle Publishing Company
One Massachusetts Avenue, NW
Washington, DC 20001

Visit us at www.regnery.com

Distributed to the trade by
National Book Network
4720-A Boston Way
Lanham, MD 20706

Printed on acid-free paper
Manufactured in the United States of America

10 9 8 7 6 5 4 3 2 1

Books are available in quantity for promotional or premium use.
Write to Director of Special Sales, Regnery Publishing, Inc., One
Massachusetts Avenue, NW, Washington, DC 20001, for information
on discounts and terms or call (202) 216-0600.

This book is dedicated to the heroes of
Flight 93. When duty called, they rose to the
task and fought for control of an airplane,
saving countless people in Washington, D.C.—
including, perhaps, me. They said, "Let's roll,"
and they made concrete the meaning
of courage and citizenship.

CONTENTS

BLAMING
AMERICA FIRST

THE THEMES AND IDEAS I WROTE ABOUT when this book was first published in April 2002* have surfaced, once again, in the debate over a prospective military campaign in Iraq—a campaign for regime change, a campaign for the liberation of the Iraqi people from the grip of a vicious madman, and a campaign to eliminate the weapons of mass destruction in that madman's hands. We have engaged in this great debate because we are the only nation that can lead such a campaign, and because we unfortunately did not complete the job of bringing freedom to the Iraqi people eleven years ago, during the liberation of Kuwait. It is a remarkable place to

---

* This is an updated and expanded version of *Why We Fight: Moral Clarity and the War on Terrorism*, Doubleday, 2002.

be: A war has to be fought, and we are the only ones who can lead it.

A year ago, as I was putting the final touches on the original manuscript of this book and initiating a project called Americans for Victory Over Terrorism (www.avot.org), I predicted that an antiwar movement would arise in our country, and that it would be animated not only by reasonable arguments about the proper use of force but also, more perniciously, by a jejune pacifism and a reflexive anti-Americanism. Some critics immediately responded that I was erecting a straw man—that no dissent from American policy was on the horizon. "[T]hese days, there is no anti-war ferment in America," wrote Walter Shapiro of *USA Today*, "aside from a handful of college students who naively believe that world peace can be achieved through sugar-free bake sales." To clinch the point, Shapiro cited poll data: "91% of Americans approve of our military action against terrorism."

My argument, however, was that this support would weaken over time—especially as the pseudo-sophisticates and the pacifists tested, honed, and marshaled their positions and their resources, and as the war against terrorism began to take on larger and costlier dimensions. And now, one year later, something very much like an intense antiwar movement is indeed strong and growing—and it is not restricted to "a handful of college students."

Late last spring, a group called Not in Our Name, whose signatories include popular entertainers and intellectuals like Ed Asner, Noam Chomsky, Ossie Davis, Casey Kasem, Martin Luther King III, Barbara Kingsolver, Rabbi Michael Lerner,

Gloria Steinem, Alice Walker, and others, publicized a "Statement of Conscience." It declared, among other things: "We believe that people of conscience must take responsibility for what their own governments do—we must first of all oppose the injustice that is done in our own name. Thus we call on all Americans to resist the war and repression that has been loosed on the world by the Bush administration. It is unjust, immoral, and illegitimate."

This was written in June 2002. On October 6, the organization, apparently undeterred by the "repression" imposed by the Bush administration, sponsored a rally in New York City that attracted upwards of twenty thousand people.

On a number of college campuses, antiwar sentiments have focused specifically against Israel, America's stalwart friend and principal ally in the Middle East. In a conscious echo of the campaigns in the early 1980s against the apartheid regime of South Africa, universities are being petitioned to "disinvest" from Israel—as if that vibrantly democratic and sorely beset land were the incarnation of racist evil instead of a country where any citizen of any race or religion can vote, can serve in elected office, and can speak out in perfect freedom. These antiwar professors and students campaign against Israel, but utter not a peep of protest against Middle Eastern regimes where Jews are not even permitted to hold citizenship or own property. Nor do these anti-Israel students and professors condemn the routine torture of dissidents in countries like Jordan and Egypt, let alone against the horror that is Iraq. The antiwar, anti-Israel movement has little to do with peace and democracy. It has everything to do with blaming America and America's allies first.

Here is another manifestation of the "antiwar ferment" that I was assured did not and would not exist: according to an October 13 story in the *Washington Post*, "The Institute for Policy Studies, a liberal Washington think tank, had compiled a list of more than 250 anti-war events planned throughout the country over the next two weeks, only to discover it had missed at least 150 others. 'People are organizing at all levels,' said Amy Quinn, co-director of the institute. 'I'm hearing from the older generations that there was nowhere near this level of activism at this stage in the Vietnam war. I'm not surprised that people are coming out against the war. I am surprised at how organized and vocal people are.'"

David Corn, the respected Washington editor of the left-wing *Nation* magazine, authored a November article in the *LA Weekly* detailing a major rally in Washington, D.C., under the auspices of an organization called ANSWER (Act Now to Stop War and End Racism). As Corn himself observed, "war" was only an excuse for this rally—which featured speeches to free H. Rap Brown* and Leonard Peltier[†]—and was assembled, among other things, to "fight against Zionism." The ANSWER rally, as Corn put it, "was essentially organized by the Workers World Party, a small political sect that years ago split from the Socialist Workers Party to support the Soviet invasion of

---

*Not your typical man of peace, Brown is a former Black Panther who also goes by his Muslim name, Jamil Abdullah Al-Amin, and is serving a life sentence for murdering one police officer and wounding another when they tried to serve him with an arrest warrant in 2000.

[†]Peltier is a leader of the American Indian Movement who is serving two life sentences for the killing of two FBI agents in 1975.

Hungary in 1956." In other words, this Washington rally was an occasion for reviving the fortunes of a group of erstwhile sympathizers of the Soviet Union—the scourge of another era—and of haters of democracy. And yet it attracted a crowd of "tens of thousands" and respectful coverage in the press.

Even some elements of the business community have joined in. In a full-page advertisement in the *New York Times*, the name of Ben Cohen of Ben & Jerry's led off a long list of signatories affiliated with well-known companies like Warner Brothers, Simon & Schuster, Rockport, A. G. Edwards & Sons, and others. This congeries of accomplished and successful captains of industry and free enterprise, speaking as "seasoned business people," asked the president of the United States, "How can blowing up buildings and killing people be good for business, unless it's the body-bag business?" Although these people had presumably neglected to poll the tens of millions suffering under the heel of the Iraqi dictator, they further asserted, in words taken from our own Declaration of Independence, that "Far from showing a 'decent respect for the opinions of mankind . . . the President is showing his contempt."

It is interesting—it is instructive—that whenever and wherever oppressed people risk their lives for freedom, as we saw most memorably in Tiananmen Square in China, they carry copies of the Declaration of Independence and homemade Statues of Liberty. Such people look desperately to America for support and help. Is it out of "contempt" for their opinions that in the war on terror we propose to unshackle the people of Iraq—as we did the people of Afghanistan—through the exercise of American power? Can "seasoned business people" really believe that such an exercise would be undertaken solely for the good of the "body-bag business"? Or is this grotesque idea a means of attracting customers to another business altogether—the hate-America business—which seems to be thriving among us once again?

Pacifists are wont to cite Benjamin Franklin's dictum that "There never was a good war or a bad peace." But Franklin wrote these words even as he was marshaling support for our own war of independence. Peter Beinart, the editor of the *New Republic*, formulated a better, and contemporary, version of Franklin's thought on CNN as his closing message for the year 2001: "There are some peaces that are worse than war. There are some things that are worth fighting for. There are some wars that you need to fight to have a just peace in the end."

This lesson has not been widely taught among us in a very long time, and in blame-America-first precincts it has not been heard in an even longer time. As the United States routed the Taliban and al-Qaeda forces from Afghanistan— "bombing a country back *out* of the Stone Age," as Christopher Hitchens well put it—the antiwar movement predicted a Vietnam-like quagmire or an American defeat on the model of the Russian experience in Afghanistan decades earlier. As the United States hunted down terrorist cells around the world, the self-described "peace movement" responded with protests and marches. And as President Bush and his national-security team highlighted the incorrigible menace of Saddam Hussein, the blame-America-firsters collected signatures and published advertising copy against removing the tyrannical regime from Iraq.

One reads the statements of the blame-America-firsters, and listens to their chants, and one wonders if they have ever absorbed a single fact about the despicable character of Saddam Hussein, or about the murder, rape, and pillage that he has inflicted on his own and neighboring peoples. If some of them have indeed perused the voluminous and scrupulously documented materials prepared by governments as well as by human-rights and intelligence organizations, and if they yet continue to hold to their position, one can only conclude that, in their view, it is better that innocent people suffer, and that a dictator amass and threaten to use horrific weapons of destruction, than that we act against such evils. For them, slavery is preferable to freedom, if that freedom comes at the price of military action by the United States.

There is a name for this attitude, and the name, once again, is anti-Americanism. In today's circumstances, in the face of an evil that only the United States can defeat, to hold this attitude is worse than irresponsible; it is a species of deep perversity. From the rest of us it requires a renewed response, one based on a true knowledge of our enemies and, especially, on a true knowledge of ourselves. This book is intended as a contribution to such knowledge.

—New Year's Day, 2003

A MOMENT
OF CLARITY

IN THE CITY WHERE I LIVE, Washington, D.C., the cleanup
was well underway by late September 2001, scant weeks after
that bright Tuesday morning when men whom none of us
had ever heard of hurled airliners into the twin towers of the
World Trade Center and the Pentagon, and jolted us into a
new and violent reality. The president of the United States
had found his voice, speaking eloquently and compellingly,
bringing the nation together to mourn, but also to fight. The
secretary of defense was directing the preparation for that
fight, readying our armed forces to strike back at those who
had attacked us. Meanwhile, in New York City, the cleanup at
the World Trade Center remained a daunting task; firemen,
police officers, and volunteers from around the country had

formed a powerful team, working together under the leadership of Mayor Rudy Giuliani to repair and rebuild the greatest city in the world.

And elsewhere? Across the nation, patriotic ardor burned bright. Suddenly flags were flying everywhere, and everywhere we were singing the national anthem and "America, the Beautiful." Charitable donations and volunteerism were only the outward, visible signs of an inward wave of sympathy and solace for those who had lost loved ones on September 11. Righteous anger and resolve had joined in support of our leaders, our armed forces, our country. For the first time in a long while there was a palpable, shared sense that this was indeed our country, and that it was a country worth fighting for.

In the wake of September 11, the doubts and questions that had only recently plagued Americans about their nation seemed to fade into insignificance. Good was distinguished from evil, truth from falsehood. We were firm, dedicated, unified. It was, in short, a moment of moral clarity—a moment when we began to rediscover ourselves as one people even as we began to gird for battle with a not yet fully defined foe.

As someone who had done his share of worrying about the moral disposition of the American people in the 1990s, I was encouraged by what I was witnessing. But moments of moral clarity are rare in life, and they are exceedingly precious. They usually follow upon hours—years—of moral confusion; they seldom arrive all at once or definitively; and they are never accompanied by a lifetime guarantee. Could

this one be trusted? It seemed we had truly begun to rediscover ourselves. But given where we were coming from, the
voyage was bound to be lengthy and arduous, the route
strewn with false turnings.

As if to remind myself of the dangers, I picked up the
paper one morning and read how, in New York City, a journalist had approached a cluster of young people lounging in
the sunshine in Washington Square Park, many of them students at New York University. They were within walking distance of the still-smoking ruins of the World Trade Center.
Asked by the reporter whether they would consider taking up
arms to defend their country and their civilization against
those who, only two and a half weeks earlier, had incinerat-
ed to death thousands of their fellow Americans, each, in his
or her own way, demurred. One said he was unwilling to
endanger his personal hopes of becoming a filmmaker:
"There are," he opined, plenty of "people who are more willing to fight, who have the mind-set of killing." Another
objected that "we're not about causes here. We're about individualism." A third, trumping the other two, offered: "This is
all [America's] fault anyway."

I'll admit that it made me a little angry, and the worst of
it, as I well knew, was that this was hardly the only example
of its kind. In the pages of newspapers, on television talk
shows and call-in radio, in Internet chat rooms, in the weekly opinion magazines, in the intellectual journals, in the
United States and in Europe and around the world, what
happened on September 11, 2001, why it happened, and
what should be done about it were the stuff of endless dis-

cussion. And in that discussion, individuals expressing views like those of the young people in Washington Square Park, and many more voicing attitudes and arguments along similar lines, with greater or lesser sophistication, occupied one highly visible corner. However genuine might be their feelings of sympathy for the victims, however appalled they might be by what occurred that day, however persuaded that "justice" needed to be served, their response was nevertheless conditioned by the fact that they were, more or less habitually, skeptical if not disdainful of American purposes in the world and reflexively unprepared to rally to America's side.

Some of them were filled with love, some of them were filled with hate, and some were merely confused:

> Force will get us nowhere. It is reparations that are owing, not retribution.

> I firmly believe the only punishment that works is love.

> Where is the acknowledgment that this was not a "cowardly" attack on "civilization" or "liberty" or "humanity" or "the free world" but an attack on the world's self-proclaimed superpower, undertaken as a consequence of specific American alliances and actions? How many citizens are aware of the ongoing American bombing of Iraq? And if the word *cowardly* is to be used, it might be more aptly applied to those [Americans] who kill from beyond the range of retaliation, high in the sky, than to those willing to die themselves in order to kill others.

We all know that one man's terrorist is another man's freedom fighter.

America, America. What did you do—either intentionally or unintentionally—in the world order, in Central America, in Africa where bombs are still blasting?

U.S. foreign policy is soaked in blood.

What is Osama bin Laden? He's America's family secret. . . . The savage twin of all that purports to be beautiful and civilized. . . . Now that the family secret has been spilled, the twins are blurring into one another and becoming interchangeable.

The World Trade Center disaster is a globalized version of the Columbine High School disaster. When you bully people long enough, they are going to strike back.

How we dare even prate about democracy is beyond me.

And so forth and so on. Of course, getting angry was no answer. Nor was it a sufficient response to mock the pretensions of such people, though I was sorely tempted to do so. After all, were it not for the vigilant exercise of American power, and the readiness of Americans to fight and die for liberty, the "individualism" cherished by those New York City college students would be as dead a letter in Washington Square Park as it was in Kabul under the Taliban. Now here they were, basking in a moral luxury bought and paid for by the object of their contempt. That America's critics everywhere in the democratic West could speak of its faults without fear and without check—was this not itself a function, an artifact, of rights won and secured by America, of safety underwritten by America, of consciences set at ease by the beneficence of America?

But the real issue was how widespread such views were, and how connected they were with the feelings of other Americans, and what effect they could be expected to have on our newfound moral clarity. To judge from one perspective— the public-opinion polls supporting the president and the war effort, the flags, the anthems—they were not widespread at all. But influence was another matter. Yes, the majority of Americans had achieved, or perhaps never entirely lost, moral clarity about our nation. But how secure was it? And would it last?

THESE WERE NOT IDLE QUESTIONS. For the truth is that we were *all* caught unprepared by September 11.

By "we" I do not mean our government, our military, or our diplomats. Of course, they, too, had been caught unprepared. Much would soon be written about their terrible omissions, ranging from severe defects in intelligence-gathering and evaluation to our overly permissive immigration procedures, faulty policing, and the indulgent and self-deceiving posture we had assumed toward the world, particularly those who hated us, in the decade after the collapse of

Soviet communism and the end of the cold war. On the issue of our physical vulnerability, I was reminded in those days of the prescient words of James Madison, warning in Federalist 41 about the "terrors of conflagration" and in particular about the dangerously exposed situation of the island of Manhattan, a "great reservoir of . . . wealth" that might be regarded "as a hostage for ignominious compliances with the dictates of a foreign enemy or even with the rapacious demands of pirates and barbarians."

All of this was highly relevant to our present situation. But when I say "we," I have in mind not so much the institutions of government as the rest of us: we, the citizens of the United States. And when I say that we were unprepared I am speaking not of physical unpreparedness but of intellectual and moral unpreparedness.

How was it that, in the wake of the bloodiest and most devastating attack on American citizens in our history, sensible and patriotic people could ask, *"Did we bring this on ourselves, by the way we have behaved in the world?"* Or, *"If we go to war against them, does that make us as bad as they are?"* Or, *"Shouldn't we work on getting rid of the poverty and oppression that are the root causes of terrorism, instead of just adding to the killing?"*

Such questions were hardly unanswerable; indeed, I mean to answer them in the course of this book. But that they could have been asked in all innocence, and that they should have been the *first* questions some of us asked, bespoke a deep ignorance not only about the rest of the world but more urgently and much more disturbingly about

America. And it bespoke an even deeper want of clarity about the difference between good and evil. September 11 had seemed to dispel this lack of clarity, or at least so I wanted to believe. But even those of us who were singing the loudest were asking such questions, the answers to which had been lost in the moral murkiness of, so to speak, September 10.

How had this lack of clarity come about in the first place? In many cases it reflected, in blander and more acceptable form, a well-developed, well-entrenched judgment about our country, and about the democratic West in general, that had come to dominate virtually every one of our major cultural and educational institutions. In time, this same adverse judgment had made itself felt in the opinions expressed in our leading newspapers, in the sermons preached in our churches and synagogues, in the causes supported by our major philanthropic institutions, in the positions on public issues espoused by the heads of our largest corporations, and everywhere in our politics.

I have called this judgment adverse. It was also perverse. For forty years, leading educators and intellectuals had been saying and writing and teaching that the United States was no better and might even be worse than its enemies; that Western "civilization," sometimes mockingly put in quotation marks, was a mask under which one perfidy after another had been visited upon the poorer nations of the world; that good and evil themselves were matters of perspective, if not of mere opinion. Some of the noblest ideas ever framed by the mind of man, including democracy, patriotism, honor, and freedom, had been systematically drained of meaning; to some younger Americans, they were now without content altogether.

The result, I would argue, was to sow a truly widespread and debilitating confusion as to our most basic national purposes, a confusion that was expressed in various forms in the wake of September 11. It was expressed by public figures who tended automatically to categorize the attacks on the World Trade Center and the Pentagon not as acts of war but as a species of natural disaster, requiring unstinting generosity toward the victims, perhaps even some limited police action abroad, but not any arousal of the American national will. It was expressed by *other* college students—we will meet them, too, in the pages of this book—who actually wanted to show solidarity with their country but found themselves inhibited to the point of speechlessness by what they had been taught about its fundamental inequities and flaws. And it was expressed to perfection in the innocent questions I cited just a moment ago about our very legitimacy.

Why, I wondered, were not more of us angry? Why did so many, especially in the country's elite, seem to back away from any hint of righteous anger as if it were some kind of poisonous snake? In the national media, anger was discouraged, denigrated, even mocked. God forbid we should act out of anger, or express a sense of righteous indignation. This raised a question in my own mind: How had such a vocal and influential minority of Americans come to believe that, to be moral, we must disavow and rise above our anger? Why wasn't anger itself considered a moral response to unprovoked attack? And why did so many people who really were angry hesitate to say so, as if it were something shameful?

That inhibition was well exploited by those whose characteristic response to the events of September 11 could be captured in the disabling phrase "Yes, but . . ." Yes, they would concede, what happened on September 11 was to be condemned in no uncertain terms, but (to quote the director of a center for constitutional rights) "people feel that there must be an alternative policy, that war cannot be the only answer." Yes, we had been attacked, but any classroom discussion of those attacks (according to a curriculum guide at Brown University) must include a call "for understanding why people resent the United States." Yes, terrorism should be eliminated, but the way to combat it (declared a scholar of international law) was by "arrest[ing] the criminals with the goal of achieving justice, not revenge," and by working to ameliorate the "cultures of poverty, oppression, and ignorance" in which terrorist impulses are supposedly bred. Yes, we had to do *something*, but in beginning to act (wrote an op-ed columnist) we were inciting jingoism, trampling civil liberties, and encouraging a "nationalist undertow that is culturally conformist, ethnically exclusive, and belligerently militaristic"—becoming, in short, just like our declared enemies. Yes, a terrible wound was inflicted upon us, but (according to a concerned citizen writing to a local paper) the courage we needed was not the courage to avenge the wound but "the courage and honesty to search our souls and recognize our [own] wrongdoings."

Why do they hate us? This was the earnest and finally irrelevant question many of us put to ourselves after

September 11. For a while it seemed to emblazon itself on the cover of every newsweekly, to occupy every nightly discussion on television. Why do they hate us? As if, by sympathetically attempting to see things from their point of view, we might figure out some way to satisfy their grievances, or assuage them with an apology. For if they hate us, the reasoning went, it must be—mustn't it?—because of something we had done to them.

I shall have more to say about this peculiarly damaging habit of mind, in which self-abasement and deference to the viewpoint of the "other" mixed unpleasantly with a new and deadly form of American hubris. Here let me just observe that if the "they" in question were extremist Muslims of the Osama bin Laden type, it was really not very hard to understand why they hated us. They had been more than forthcoming about their sentiments for a very long time, and after September 11, bin Laden himself was permitted to expound the theme at length on our home screens. Their vicious rage was not obscure; neither, as they had made spectacularly clear by their self-immolating actions, was it to be appeased.

For some of us, at any rate, it seemed to have become a harder task to extend to *our* values and *our* purposes the same degree of open-minded sympathy we were ready to extend to theirs. It was harder still, evidently, to muster a stouthearted response to their implacable intentions. "Don't let hate grow in our hearts" read a sign a friend of mine noticed in a hardware store soon after September 11. It made a perfect counterpart to "Why do they hate us?" In both there was the same unwillingness to credit the objective reality of our situation, the same wishful notion that the problem we faced lay wholly within ourselves, in our emotions and attitudes, and that the solution therefore lay in an adjustment of those attitudes.

Well, I had not seen much hate in American hearts—a bit, very copiously documented in the press in the form of scattered bias incidents against Muslim Americans, but not much. What I was seeing instead was either a reluctance or an inability to find words, ideas, arguments, rhetoric, or models adequate to the gravity of the crisis and to the heroic scale of commitment that was needed to face and overcome it.

Of one thing I was sure: The critics of American purposes would not cease their work. I was also sure of something else: Their sentiments, when framed as criticisms of American purposes, were not shared by large numbers of people. But neither were those sentiments being answered, either because the answers to them had been forgotten or because the answers had never been learned in the first place. I sensed in my bones that if we could not find a way to justify our patriotic instincts, and to answer the arguments of those who did not share them, we would be undone.

And that is when I resolved to write this book. *Why We Fight* is my effort to answer the questions being asked about this war—not only in the immediate aftermath of September 11 but as our counterattack got underway and to this very day—questions that can only intensify in the coming period as the battle widens and opposition to it grows and becomes more articulate. *Were* we justified in replying to force with force, or should we have pursued another route, especially the route of international law? Is our culture "better" than others, and on what moral and intellectual grounds can it be defended? Why *do* they hate us—and who exactly are "they," whom do they represent, and what do they stand for? Were we dragged into this war by our "one-sided" support for Israel, and what is the rationale behind our friendship with that country? Is there something suspect—something jingoistic—about plain old-fashioned patriotism?

Moral questions all, and my answers are likewise framed largely in moral terms. This is not, in other words, a book about policy. I have not tackled the issue of our military strategy, though anyone attending to my arguments can readily deduce that I wholeheartedly advocate pursuing this war to final victory. Nor do I say anything about what our immigration policy should be, though it is well known that in general I am both pro-immigrant and fully in favor of strong steps to control our borders. I touch only lightly on issues of civil liberties, though I would defend both the idea of military tribunals for terrorists accused of war crimes and the detention of suspects within our own borders for questioning. The matters dealt with here are mostly of a different order; their connection with policy is implicit, not explicit, and I have deliberately refrained from making concrete recommendations.

Throughout this book I have tried to differentiate between the doubts held by well-meaning Americans and the arguments of the critics who feed those doubts. To the latter I have tried to respond fairly, and at their strongest rather than at their weakest points. I also hope that I have not ceded any ground. For there is, in truth, much to be angry about in the positions some of these critics espouse: much that is meretricious, much that is inspired less by reason or evidence than by simple ideology and, yes, moral perversity. They have caused damage, and they need to be held to account.

The damage is to be measured in our loss of memory. In one way or another, there has been a great forgetting, and the result has been a kind of unilateral disarmament. Now we have been caught with our defenses down—our intellectual and moral defenses as much as our physical ones. Many of us have allowed ourselves to forget what we once knew about this country, which Lincoln called the "last, best hope of earth," and we have forgotten why he was right to call it that. Many of us have forgotten what we once knew about our unique institutions of government, and we have neglected to implant in the hearts of our young people the never-ending duty to preserve and protect them. Many of us have forgotten what we once knew about our freedoms and our decencies, and we have forgotten why, time and time again, we have had to rally ourselves to the point of ultimate sacrifice to defend them. Many of us have forgotten the truth we once knew about the heritage of our Western civilization, and we have all but forgotten where we put the key to that truth, or whether we have a right to reclaim it.

The time has come to begin remembering, and to rearm.

—New Year's Day, 2002

CHAPTER 1

THE MORALITY OF ANGER

THE RUINS OF THE WORLD TRADE Center were still smoking, ash and soot lingered in the air, the odor of death lay everywhere. It was early October 2001, and one army—an army of police and firefighters and rescue workers and volunteers of every stripe—was hard at work clearing, searching, burying, shifting mortar, ministering to mortals. Another army, under the direction of the president and the secretary of defense, was readying itself to move against our attackers. The land was full of grief and full of anger, and full of opinion.

What had happened to us? What could we do about it? What *should* we do about it?

We were not the only ones asking. In the days after September 11, the whole world caught its breath, waiting to see how we would respond. Ordinary people everywhere shared our shock and astonishment, sympathized with our grief, understood our anger, were moved by our unity and solidarity. But both at home and abroad there was also uncertainty, even apprehension, as to what we were going to do about this assault. Would our response be measured and appropriate, or would we strike out blindly, thereby confirming the lowest expectations of both foreign and domestic elite opinion? Long before we responded, the nature of our response had become, for many, a test of our national character.

From where I sat, the quality of both the grief and the anger—fierce, aroused, yet deeply thoughtful—was a sign of everything that is instinctually grand about the American national character. I had agonized for years about what was happening to this American character as our educational standards spiraled ever downward, our elites presided over an unprecedented coarsening of our culture, and our people seemed to be showing clear signs of self-doubt and moral confusion. The truth is that I would rather have gone on agonizing forever than have had my questions answered by a national calamity, but when that calamity occurred on September 11, the overwhelming and immediate reaction of our people—not the grief and anger in themselves but the *quality* of the grief and the anger—certainly helped to answer them.

As for the quality of post–September 11 opinion, on the whole it, too, bespoke the settled maturity of the American people, tending as it did to coalesce around a consensus view that retaliation had to be swift and uncompromising, adequate to the outrage, and in keeping with the dictates of our moral and political traditions. But there were other opinions as well, motivated, primarily, by the fear that we would over-react, that September 11 would trigger our supposed tenden-cy to blind rage and rash action. Suddenly the name of Curtis LeMay, the American general who was alleged to have rec-ommended that we "bomb Vietnam back to the Stone Age," was in the air again, a code word for what was assumed to be the "default" mode of American military thinking.

In fact, those among us who espoused the LeMay position were scarcely to be heard from. By contrast, what might be called the Gandhi position—the position of nonviolence—was articulated in many places, was treated with exceptional seriousness by the media, and was amplified accordingly. It was also amazingly quick to materialize. Indeed, the opera-tions of our domestic "peace party" gave fresh meaning to the Coast Guard motto *Semper Paratus*. Without benefit of a cen-tral command, without training manuals, without field exer-cises, it was able to deploy its forces with lightning speed, to seize the attention of the press, and to read from a single script. Its tactics—and its instincts—were models of rapid mobilization.

"I don't think the solution to violence is more violence," opined a Columbia University sophomore to a reporter as she held up a sign—"Amerika! Get a Clue!"—at an antiwar rally in Washington in late September. Said a mother in Kennebunk, Maine, around the same time: "Killing people won't prove anything. It's just more of the same." At a protest demonstration in early October in New York City, just blocks from the smoldering ruins of the World Trade Center, Ronald

Daniels of the Center for Constitutional Rights asserted with confidence that "war cannot be the only answer" and pleaded for "an alternative policy." In San Francisco, an advocate of women's rights blamed the media for "whipp[ing] up to a great extent the call for vengeance or the call for war." In Wisconsin, a protestor lamented "all the flags out supporting the slaughter."

Most of these events were held long prior to anything we had done or even talked about doing in response to September 11. They reflected, rather, a deeply held prejudice about the proper way to deal with conflict and aggression, and an equally deep mistrust of the good faith of the American government.

Some in the peace party were already going farther, shifting the subject away from the attack itself and toward the behavior—past, present, or future—of the United States. At the New York City demonstration, a representative of Vietnam Veterans Against War told the crowd he did not "want to see more Americans die because of a militarist cowboy"—the militarist he had in mind being not Osama bin Laden but the president of the United States. A professor at Brown University instructed his audience that if "what happened on September 11 was terrorism," what America had done "during the Gulf War was also terrorism." Such sentiments were echoed around the world, in places as diverse as Canada ("shut down the American war machine") and Athens, Greece, where four thousand people marched in opposition to an "imperialist war" started by "Americans, murderers of peoples."

As the weeks wore on, admittedly, pronouncements of this kind did tend to wane in intensity. How could they not? The military campaign in Afghanistan was planned so scrupulously and conducted with such care, achieved such a stunning success so quickly, with so little loss of innocent life, and to such unmixed joy among the Afghan people, that the edge of protest was blunted. Even on university campuses, antiwar sentiment faded and pro-war and pro-American sentiment became tolerable if perhaps not yet fully respectable. Many students, though many fewer professors, actually discovered the morality of military action.

If, then, the nature of our response was a test of our national character, it was one we would seem to have passed with flying colors. Or so things stood at the turn of the year 2002. But even then, in the interlude after the fall of the Taliban, it was clear that all this could change once more. For the larger, global war against terrorism was far from over, and from here on in, things were only likely to get more complicated. India and Pakistan, two of our partners, were already at each other's throats. The great question of whether we were going to go after Saddam Hussein hung before us. No solution to the Israel-Palestinian conflict seemed in sight, and to some it was beginning to seem that Yasir Arafat's Palestinian Authority should itself be placed on the list of terror suspects.

In short, military campaigns were almost certainly bound to become tougher and more protracted in the period ahead. This in turn suggested that coalition partners might break away, and world opinion might shift. American forces could begin to take significant casualties; there might be mounting concerns about civil liberties at home; the domestic consensus might weaken, thus endangering success in the war.

Weakening that consensus, sowing and reinforcing doubt about our purposes and our methods, was in fact the goal of the peace party. Its favored means: casting a shadow of moral doubt over our righteous and justified anger, promoting the idea that our tendency to jingoistic aggression could only be checked by a countercommitment to nonviolence. The celerity with which the peace party proved able to mobilize and make itself felt, in the face of an unambiguous and monstrous aggression on our soil, suggests not only the deep-rootedness of its own attitudes but the potentially wider effect those attitudes might yet have on national morale.

In later chapters we will deal with the workings of some of these same attitudes in relation to such issues as cultural confidence and love of country. Here I want to focus more narrowly on war and peace, force and pacifism. By looking at the national debate over these matters in the early days of phase one of our war, we can learn important lessons for phase two and all the phases to come. For the arguments are not going to go away.

I MENTIONED THE WORD *pacifism*, and right away I need to make a distinction. There is such a thing as a genuine predisposition against violence in human affairs, and it has roots in very old traditions of thought. There is also a particular version of this orientation that has its origins in more recent doctrines, including certain psychological theories about the role of "aggression" in men and boys. And then there is a form of pacifism that is disposed not so much against the use of military force in general as against the use of military force by one particular actor, the United States of America. This last-named type of pacifism derives from a negative view of the ends for which American force has allegedly been exercised in the past, or from a more free-floating hostility to America as a society—or both.

The strands are also often conjoined, with a seemingly principled pacifism serving as a "cover" for anti-Americanism. Thus, the Columbia University student who declared that violence is no solution was holding a sign on which the sixties-style spelling "Amerika" was meant to suggest a parallel between this country and Nazi Germany. Or take the instruction imparted to its young charges by the Mount Rainier Elementary School just outside Washington, D.C., which sees its "most important responsibility," according to a report in the *Washington Post*, as ensuring that there will be "no fighting." Here is the catechism as filtered through the sensibility of one eleven-year-old boy: "I believe in peace—in not fighting and treating people with respect. . . . We learned in our class that if you believe in peace, you can stay alive. We learned that you should always find a peaceful way to solve your problems because you should never be violent."

Why do you suppose this boy's teachers sought to drive home their dreamy message after September 11? Certainly, in my view, not in order to chastise the violent men who sought to solve *their* "problems" by massacring our civilians. Rather, I suspect, they were seeking to deliver a preemptive judgment against the president, to prevent another generation of young people from learning the proper uses of righteous anger, and to throw dust in the eyes of the American people.

I will return to the lessons being taught by schools like this one, but first I want to take up the older and politically untainted traditions of pacifist thought to which I alluded, and try to give them their due. I have in mind the traditions connected with religious teachings, and specifically with Christianity. In the West, Judaism has produced its own rich writings on violence and war, but in contrast with Christianity or at least with Catholicism, religious authority in Judaism has never resided in a central body, and there is nothing in it corresponding to Church dogma; besides, the two-thousand-year historical experience of the Jewish people from the end of the biblical period to the founding of the state of Israel was the experience of a minority lacking sovereign power or the means to deploy military force. Islam, by contrast, was a religion connected with power and conquest from the beginning, and as we shall see in Chapter 3, pacifism in the usual sense is quite alien to it.

Christianity, too, developed in relation to earthly power, but that relation, at least in the early centuries, was oppositional; Christians were but a small persecuted sect on the fringes of the Roman Empire. Yet even after this ceased to be the case and Christianity became the official religion of the empire, the influence of certain seminal passages about peace in the teachings of Jesus never waned. So it is no surprise that perhaps the most eloquent and passionate defenders of pacifism today are those Christians who, appealing to the New Testament, hold that violence is never justified or justifiable, and that the injunction to turn the other cheek admits of no exceptions.

These people—they include such groups as Anabaptists, Mennonites, Quakers, the Amish, but also individuals and organizations from more mainstream denominations, both Protestant and Catholic—believe quite sincerely in the principles of nonviolence. The integrity with which they have striven to maintain those principles is admirable. Although they received relatively little attention in the post–September 11 period, they have hardly flinched from making known their convictions—and, in some cases, their honest struggles with those convictions.

Thus, to the Reverend Graylan Hagler of the Plymouth Congregational Church in Washington, D.C., it was a given that our response to violence must not be a military one— for, as Jesus taught, "Blessed are the peacemakers" (Matthew 5:9). If we were to choose the road of violence, warned the Reverend Hagler, "the reaction [would] only be [more] violence. . . . In a world of an eye for an eye, a tooth for a tooth, the world ends up blind and toothless." Similarly, to Ed Crayton, a black man who grew up in the Baptist and Lutheran churches, "There's nothing in the Bible that talks about . . . Jesus giving us a chance to wage war." A letter writer contributed this to the *Washington Post*: "The message of Jesus Christ is the ultimate solution to the conflagration.

. . . We must categorically renounce violence as an instrument of international activity." Remember, admonished a reader of the *Atlanta Journal-Constitution*, "Christ was an absolute pacifist."

Others were more troubled, clearly caught in an inner struggle between their religious conscience and their instinct for justice. Tom Robert in the *National Catholic Reporter* wrote of "trying like crazy to wiggle out from under . . . the difficult sayings about non-violence that keep creeping out of the story of Jesus of Nazareth." To Julie Ryan of the Dallas Peace Center, the "ultimate challenge" at this moment was to hew to the teaching of Jesus to "love our enemies." Father Matthew Ruhl, a Jesuit pastor in Kansas City, identified the injunction to turn the other cheek as "one of the most distressing teachings of Jesus for me." But in the end, these, too, accepted pacifism as a defining tenet of Christian faith.

Once again, sentiments like these tended to fade as the campaign in Afghanistan got underway, and especially as it became clear that we were taking extraordinary steps to avoid civilian casualties. Still, in mid-December, an ad hoc coalition of sixty-eight Catholic organizations and individuals called on the Catholic Church to denounce the war in Afghanistan as immoral and in violation of religious doctrine. In so doing, they were defying the position of the U.S. Conference of Catholic Bishops, which shortly after September 11 had declared forthrightly in a letter to President Bush that "our nation . . . has a moral right and a grave obligation to defend the common good against such terrorist attacks." Whether or not this new dissent from the

Church's position was a portent of things to come, it raised in stark form the question of whether the Christian tradition does in fact pose a principled objection to war, and in particular to this war, that we must take seriously.

LET US GO BACK TO THE PHRASE "an eye for an eye, a tooth for a tooth." It comes from the book of Exodus, and it was indeed an idea to which Jesus took explicit and quite radical exception. Going beyond the by-then conventional understanding of the rabbis, which taught that the phrase was not to be taken literally but as a demand for appropriate monetary compensation (and no more than appropriate monetary compensation), Jesus negated the idea of compensation altogether, let alone retribution: "I say to you, Do not resist an evildoer. If anyone strikes you on the right cheek, turn the other also." More: "I say to you, Love your enemies, and pray for those who persecute you, so that you may be children of your Father in heaven." In the garden of Gethsemane, according to Matthew, Jesus even rebuked Peter for drawing his sword and striking a slave of the high priest in an attempt to prevent the arrest of Jesus: "Put your sword back into its place, for all who take the sword will perish by the sword."

These famous texts are straightforward enough, and in their unequivocal aversion to the use of force they have resonated down the centuries with a clarion purity. But as with so much else in the Bible, they are not the only or the last word on the matter; they are not even Jesus' own last words on the matter. In one of his few unmixed utterances of praise, for example, Jesus lauded the faith of a Roman centurion, a soldier and a man of violence, who, shortly after the Sermon on the Mount, approached him to heal an ailing servant; clearly, then, Jesus did not regard a belief in his own teachings, or indeed in himself, as by definition inconsistent with taking up arms.

Then there is the no less famous passage in which Jesus, though obviously speaking metaphorically, warns his disciples that he has not come to bring peace to the earth but "a sword. For I have come to set a man against his father, and a daughter against her mother . . . ; and one's foes will be members of one's own household." And let us not forget the account (in the Gospel of John) of Jesus' cleansing of the Temple in a fit of violent "zeal." ("Making a whip of cords, he drove all of them out," shouting, "'Take these things out of here! Stop making my father's house into a marketplace!'")

Even the story of Gethsemane is not so clear-cut as is often claimed. In the version given in the Gospel of Luke, Jesus not only refrains from rebuking Peter but actually urges his apostles to equip themselves with weapons. ("The one who has no sword must sell his cloak and buy one.") A still weightier complication is introduced in the Gospel of John, where Peter's infraction is defined differently. There, he is admonished not for the use of force per se but for interfering with the necessary unfolding of the divine plan: "Put your sword back into its sheath," Jesus says to Peter. "Am I not to drink the cup that the Father has given me?"

This brings us to Paul and the beginnings of the formal development of Church doctrine. Here again we encounter a somewhat mixed message. "Bless those who persecute you," the apostle writes in Romans, echoing the words of Jesus. "Do not repay evil for evil, but take thought of what is noble in the sight of all. If it is possible, live peaceably with all. Believers, never avenge yourselves." This sounds positively Gandhian in its view of violence as evil, no matter why or by whom it is committed. But immediately thereafter, Paul adjures the fledgling Church to accept and "be subject to" earthly authorities and their coercive powers, for "those authorities that exist have been instituted by God," and "the authority does not bear the sword in vain" but is rather "the servant of God to execute wrath on the wrongdoer." The same thought is repeated and embellished in the first letter of Peter, where that disciple reminds his recipients that human institutions are "sent by [God] to punish those who do wrong and praise those who do right."

Nowhere in the New Testament do we find force itself held up for explicit praise—that would be all but unthinkable. But neither are the Gospel writers so unworldly as to posit that the answer to every human conflict is to turn the other cheek; in certain circumstances and for certain purposes, force would seem to be forbidden, in other cases allowed (even if never encouraged for its own sake). Indeed, it was to elaborate the why and how and wherefore of the latter case that the Church, over the centuries, developed the doctrine of "just war," a theory that received its first extended treatment in the late fourth century at the hands of St. Augustine and was significantly modified nine centuries later by St. Thomas Aquinas.

"A great deal," Augustine wrote in *Contra Faustum*, "depends on the causes for which men undertake wars, and on the authority they have for doing so." Aquinas, specifying, named three main criteria for determining if one could initiate war. (This part of just-war doctrine was called by the medieval scholastics *jus ad bellum*, the right to go to war, as distinct from *jus in bello*, the proper conduct of war.) The three were: whether war is declared by a legitimate sovereign; whether it is for a just cause—that is, a cause that avenges wrongs or rights an injustice; and whether the belligerents "intend the advancement of good, or the avoidance of evil." To those who argued that Christians should always seek peace, Aquinas responded that those who wage war justly do, in fact, aim at true peace, being opposed only to an "evil peace." Indeed, Christians would be shirking their religious duty were they not to struggle against an unjust peace, including by taking up arms.

DID AMERICAN MILITARY ACTION in the wake of September 11 satisfy these three criteria? That it was waged by a legitimate authority is patent: that authority being the duly elected and sworn president, acting with the virtually unanimous approval of the elected representatives of the American people. Likewise, it was clearly waged in a just cause, against terrorists who sought and still seek to destroy us, as well as to avoid future evil.

True, even when a war is waged by a legitimate authority, for a legitimate reason, and for a legitimate end, other factors must weigh heavily. Implicitly referring to the tradition of *jus in bello*, the letter of the U.S. Conference of Catholic Bishops to the president warned: "Any military response must be in accord with sound moral principles, . . . such as probability of success, civilian immunity, and proportionality." Although war is certainly hell (in the pithy observation of General William Sherman), our conduct of it must nevertheless be appropriate. We may be unable to avoid injuring innocent civilians in the course of fighting, but we must not target them. Likewise, we must not kill or mistreat prisoners of war. And we must always be wary of producing, even unintentionally, evils commensurate with those we are seeking to eliminate.

Obviously there is a fine moral line here. As the scholar Jean Bethke Elshtain has pointed out, Augustine, in developing the idea of the just war, struggled with the fact that the weight of Scripture challenges the use of force. But he recognized, as most Christians have, that there are times when *not* resorting to force leads to evils far greater than the one we oppose. And as for whether we have fought a war justly (in contrast to fighting a just war), any proper assessment requires careful analysis based on specific facts—and often on the outcome of the hostilities themselves. To quote the Anglican theologian Patrick Comerford, "It is only long after a war is over that we have the time and the luxury to determine whether all conditions [of a just war] have been met."

By all these standards, both the military campaign in Afghanistan and our conduct of that campaign qualify unreservedly as just. In light of its aims and its achievements, in

consideration of our extraordinary sensitivity to the avoidance of civilian casualties, and in light of our vast efforts of
humanitarian relief for the suffering people of Afghanistan, I
would not be surprised if, in historical retrospect, the
Afghanistan campaign were to qualify as one of the most just
wars ever fought.

THE FACT THAT WAR IS ALWAYS tragic does not permit us to
compound the tragedy. But neither must it be allowed to
undo our moral compass. Even some of the most principled
advocates of religious pacifism have come to grips with the
irreducible facts of human evil and the need to oppose it by
force. One of them in the aftermath of September 11 was
Scott Simon, the well-known commentator on National
Public Radio. A Quaker himself, Simon wrote in the *Wall
Street Journal* that "about half of all draft-age Quakers enlist-
ed in World War II, believing that whatever wisdom pacifism
had to give the world, it could not defeat the murderous
schemes of Adolf Hitler and his cohorts." In the present
instance, Simon concluded, if we did not defend ourselves
and punish those who meant to destroy us, then we ourselves would be destroyed. For pacifists, he wrote, it was better "to sacrifice our ideals than to expect others to die for
them."

Perhaps no Christian thinker is more qualified to speak
to this matter than Dietrich Bonhoeffer, who did indeed sacrifice his ideals, and his life, to fight absolute evil. A Lutheran
minister in Germany, Bonhoeffer was teaching at Union

Theological Seminary in New York when World War II broke out in 1939. He promptly returned to Germany, where in time he joined a plot to assassinate Hitler. He was captured and hanged in a concentration camp in 1945.

Bonhoeffer knew the cost of Christian witness, and was prepared to pay it. His friend Reinhold Niebuhr, the American theologian and the great expositor of Christian realism, said of Bonhoeffer: "His life became for many of us a contemporary moral challenge—we looked up to him as if he'd been sent to inspire us, to serve as a reminder of what truly matters." These words should be taken to heart by every pacifist who appeals for authority to the Christian tradition.

IF I SEEM TO HAVE SPENT AN inordinate amount of time answering the points raised by religiously motivated pacifists, it is not because they are numerous among us or truly influential. Rather it is because they strike me as sincere in their beliefs, however partial or selective I find the arguments supporting those beliefs. Moreover, the arguments of principled pacifists are often invoked by those who seek to use them in support of something else. Among the latter, I am afraid, are many churchmen and -women.

The letter of the U.S. Conference of Catholic Bishops to President Bush was exactly what should be expected in a national emergency like the one we were facing. But it would be a mistake to conclude that most spokesmen of faith were suffused with the martial spirit that was once routinely associated with American Christianity. As J. Bottum reported in the *Weekly Standard* in early October, Americans who flocked to church in record numbers on the Sunday after the attacks were less likely to hear a righteous call to arms than "admonitions not to indulge in racist feelings against Arabs or to give in to the angry lust for revenge." From his observation, Bottum added, neither of these seemed to present much of a temptation: Among churchgoers, "righteousness has come to seem the equivalent of self-righteousness, and hardly anyone believes in genuinely righteous anger anymore."

This is another way of saying that many establishment churches have in effect become departments of our secular liberal culture, and to that culture we may therefore now turn.

One route to pacifism, as I mentioned earlier, runs by way of current psychological doctrine. Generations of American children have by now been raised on the principle that violence is always wrong, and that every difference can be negotiated through "dialogue." Likewise, generations of American businessmen and executives have been trained in the principles of conflict resolution and anger management. Generations of American diplomats and negotiators have been instructed in the art of "getting to yes."

What is wrong with that? Nothing—as long as the parties to a dispute are playing by the same procedural rules, as long as the matters under dispute are truly negotiable, and as long as each side can be trusted to abide by the settlement. In other circumstances, and especially in war, anger management is at best irrelevant. "Don't hit!" is easy advice; "Don't hit back!" is more fraught with complexity.

But why must we accept the premise that anger is itself a suspect quality and always in need of "management"? To the contrary, as the ancients recognized, anger is a necessary power of the soul, intimately connected with the passion for justice. The appeal to stifle our anger and negotiate our differences with extremists bent on nullifying our existence was thus not only irrelevant, it was immoral; it amounted to a counsel of unilateral disarmament and a denial of justice.

A subset of this discussion concerns the question of whether September 11 was an act of war or a crime; if it could be defined as the latter, might we not obviate the need for a violent response altogether? Our "dispute" with the perpetrators of September 11 would then be conceived in legal terms, and they themselves might be arrested and (in the words of Alan Mattlage, the organizer of the Washington protests) "tried before an international tribunal." A letter signed by Rosa Parks, the legendary heroine of the American civil rights movement, as well as by such entertainers as Harry Belafonte and Danny Glover, similarly urged that "our best chance of preventing [further] devastating attacks is to act decisively and cooperatively as part of a community of nations within the framework of international law." Once the war began, the same solicitude extended to the "rights" of Osama bin Laden was then extended by others to further potential malefactors like Saddam Hussein. All of these miscreants, it was said, belonged not in Hades but in The Hague.

In this automatic reaching for the courts and the lawbooks, and especially for the cloak of "international" jurisdiction, it seemed that matters of fundamental right and wrong were once again being elided. Mass murderers were treated like storekeepers arguing with suppliers over a shipment of mislabeled goods, and "the law" as interpreted by the United Nations was invoked as some magisterial and transcendent standard of arbitration. One would hardly know from such pious declarations that September 11 was the single most violent day in American civilian history. To reduce this mass atrocity to a "crime" against "international law" was to trivialize it beyond recognition.

IT NEVER CEASES TO AMAZE me that people who are quick to condemn their fellow Americans for cultural "chauvinism," or for failing to appreciate the sensibilities of others, should evince such narrow-minded narcissism on this point, blithely assuming that everyone else in the world accepts their idea of the best way to resolve differences. Still, I suppose I have no cause for amazement. As the history of the last years has amply demonstrated, the getting-to-yes syndrome has infected everything from our most casual domestic transactions to our conduct of foreign policy, leading us time and again to misread the malign intentions of our enemies and making us an even more inviting target for terrorists and other aggressors. It feeds into, and off of, a larger bias to which I also alluded earlier: the bias against the employment of force by the United States of America in pursuit of its interests or its national honor.

One saw this bias at work, for example, in the insistence that the war against terrorism be prosecuted by means of an international coalition and not by the United States alone. Of course, there were sound strategic reasons for securing the active cooperation of others, as we did from the start. Morally, too, there is always something to be said for having an explicit seal of approval from the global community. Something, but not everything. To make such international approval a *requirement* of action, as if otherwise we lacked warrant to defend ourselves, was not morally sound but morally repugnant, springing from a hostility to America that had little to do with pacifism and everything to do with the larger political and ideological agenda of the "peace party." The idea behind it was that we could not be trusted to restrain ourselves—an idea that no amount of evidence to the contrary could dislodge from the minds of those holding it.

For some on the Left, even the coalition so painstakingly assembled by President Bush was corrupted by the very fact that *we* had assembled it and were leading it. "What we are seeing," wrote the antiglobalist activist Naomi Klein, "is not a global response to terrorism but the internationalization of one country's foreign-policy objectives." Not only was stopping terrorism alleged to be a mere "interest" of the United States—and of the United States alone—but our pursuit of that interest was implicitly characterized as more pernicious than terrorism itself. Here, even the right to self-defense, let alone the legitimacy of acting in our national interest, was taken from us.

Softer versions of this hateful charge abounded on the Left. For Jonathan Schell, writing in the *Nation*, the fight against terrorism should have been undertaken as a highly specific "police action," with "military action play[ing] a merely supporting role—in the form, perhaps, of the occasional commando raid." The major part of our response, in Schell's judgment, should have been "a comprehensive global effort to rid the world of weapons of mass destruction," led by the "readiness of the great powers"—that is, us—"to disarm." Also in the *Nation*, Professor Richard Falk, positioning himself between two fallacies of ostensibly equal but in fact unequal weight, the "pacifist fallacy" and the "militarist fallacy," imposed so many preconditions on American military action as to render us altogether impotent.

For still others, the pacifist response hinged on a secular variant of the *jus in bello* argument: the certain knowledge that American military action would result in the death of innocents. "Killing people won't prove anything," said that mother in Maine. "It's just more of the same." Andrew Greeley, a Catholic priest and writer, added that America was "organizing its own jihad."

More of the same? Terrorists target innocent civilians by definition; they seek the destruction of innocent life. Military action to combat terrorism seeks to avoid noncombatant casualties. It is not more of the same, it is the opposite of the same. It was ludicrous to suggest that morally we would be as culpable for civilian casualties resulting from our use of force in Afghanistan as were the terrorists for the deaths of those on the planes, at the Pentagon, and in the World Trade Center. Besides, although it was unfortunately true that innocent people would die as a consequence of our action—in Afghanistan, to say it again, such casualties never approached a fraction of what was being foretold—without that action many more innocent people would *certainly* die. Moral delicacy untethered from the recognition of facts and circumstances is a form of moral idiocy.

WHAT DO WE OWE TO OUR COUNTRY, and can we allow our moral delicacy to decide *that*? One often overlooked element in the pacifist stance is its moral luxury—the fact that it is made possible, and protected, only by the willingness of others to use force. As George Orwell put it, speaking of pacifists in World War II, "Those who 'abjure' violence can only do so because others are committing violence on their behalf."

This point can be extended: In the democratic societies of the West, the critique of violence depends entirely on the continued vigilance of those who are often the prime targets of that critique, notably the armed forces. It also depends on the maintenance of a common set of expectations, too often taken for granted, regarding the norms of civilized life. In the words of the novelist and Nobel laureate V. S. Naipaul, writing about the lyrics of Joan Baez, "You couldn't listen to the sweet songs about injustice unless you expected justice and received it much of the time. You couldn't sing about the end of the world unless you felt that the world was going on and on and you were safe in it."

The same point is also relevant to the few cases in which nonviolence has actually succeeded: The obvious examples are Gandhi and Martin Luther King Jr. For Gandhi was appealing, shrewdly, to the British devotion to law and fair play, as well as to British humanitarian sentiment; were he to have adopted a similar tactic in Nazi Germany, his movement would have been brutally extinguished and his own fate would have been death. King also based his entire campaign on the well-founded certainty that most Americans were not only appalled by segregation but were unalterably committed to the universal extension of the rights and guarantees enshrined in the Constitution. In countries that respect human rights, in countries that exhibit a conscience, nonviolent protest can succeed. The idea of its succeeding with our enemies today is laughable.

It goes without saying that Islamic extremism of the kind practiced by the Taliban and espoused or nourished by others around the world leaves no room for the dissent that our democracy affords, let alone for pacifism. It also goes without saying that pacifists would not wish to live under such regimes. But their arguments, taken seriously, would prevent them—or anyone else—from doing what might be necessary to stop such regimes from arising, and would certainly save no one—including pacifists—from becoming their victims. Is that moral, or is it actually immoral? I agree with the late philosopher Sidney Hook, who wrote that absolute adherence to pacifism "makes the pacifist morally responsible for the evils that an intelligent use of force may sometimes prevent."

Today's pacifists owe their very lives to the America that stopped Hitler—nonpacifically. Were we to have accepted their moral reasoning now, we would have laid ourselves open to more grievous attack and, quite possibly, to the prospect of a world in which people holding nonviolent beliefs would be exterminated. In this sense, as C. S. Lewis prophesied, pacifism means "taking the straight road to a world in which there will be no pacifists."

WHY HAS THE CRITIQUE OF VIOLENCE taken such hold among us, and why does it exercise such influence? The answer is paradoxical. Contrary to the myth of our nation as violence-prone, Americans are in fact a peaceful people, averse to conflict. That is the larger truth about us. Our habits are the habits of a commercial society, resting on rich deposits of social trust and on laws that regulate and protect transactions of every kind. Our outlook is the outlook of a democratic polity, guided by the spirit of accommodation and compromise, superintended by guarantees of due process and judicial review.

It is exactly these same habits and this same outlook, I believe, that make so many of us susceptible to arguments of the give-peace-a-chance kind. I don't mean that we necessarily buy those arguments, but something in them appeals to something good in our nature, and though we may know they are wrong we often do not know quite how to answer them. That is precisely why I have been at such pains to take them seriously on their own terms. But I also want to be clear about where they come from.

It is theoretically true that one can espouse nonviolence *and* support the war effort. As Scott Simon reminds us, some genuine pacifists and conscientious objectors have done just that in past conflicts. The trouble with many of today's pacifists is that, in the name of the higher morality of nonviolence, they have not only declined to support the war effort but have actively tried to hamper it, loudly warning about cycles of violence, accusing us of an unseemly lust for vengeance, invoking the supposedly dark record of our past, sowing doubts about our intentions, impugning our right to defend ourselves.

In short, many in the "peace party" who cloak their arguments in moral objections to war are really expressing their hostility to America, and it does the cause of clarity no good to pretend otherwise. That hostility—in more than a few cases, *hatred* is a more accurate word—is many-sided and has a long history, and we shall be encountering facets of it in our discussion. But where armed conflict is concerned, the arguments of today's "peace party" are basically rooted in the period of the Vietnam war and its aftermath. It was then that the critique of the United States as an imperialist or "colonialist" power, wreaking its evil will on the hapless peoples of the third world, became a kind of slogan on the Left. This same critique would, in due course, find a home in certain precincts of the Democratic party and, in more diluted form, would inform the policy preferences of the Carter and Clinton administrations, and it is with us still.

It is especially prevalent in our institutions of higher learning. At a teach-in at the University of North Carolina immediately after the attack, one speaker remarked that, were he the president of the United States, his first act would be not to avenge the infamy but to apologize to "the widows and orphans, the tortured and impoverished, and all the millions of other victims of American imperialism." For a professor at Rutgers, whatever the "proximate cause" of September 11, "its ultimate cause is the fascism of U.S. foreign policy over the past decades." Like the character in Molière's *Le Bourgeois gentilhomme* who was astonished to learn that he had been speaking prose all his life, these two seemed unconscious of the fact that they were speaking clichés, and clichés with a certain identifiable provenance.

Allied to the political critique of America that developed in the 1960s and 1970s was a cultural and psychological critique. Not just imperial ambition but a sort of deranged, Wild West machismo was said to be driving our activities abroad, impelling us to drop bombs on innocent people and/or to force upon them our uniquely rapacious model of economic activity. At home, this same derangement was said to lie behind everything from our alleged obsession with guns to our alleged obsession with order and the perverse way in which we brought up our children, especially our boys. Out of this critique there arose the by-now standard counterwisdom that I have already discussed: that conflict is always a product of misunderstanding, and that violence is always wrong.

In the past decades, since Vietnam and especially since our defeat there, our culture has undergone a process that one observer has aptly termed *debellicization*. Military virtues have been devalued and shunned, and along with them the very idea that war solves anything or is ever justified. Generations of schoolchildren have been taught that conflict is something to be avoided. Parents and teachers have been cautioned by psychologists and feminists alike that male aggression is a wild and malignant force that needs to be repressed or medicated lest it burst out, as it is always on the verge of doing, in murderous behavior. The 1999 shooting spree by two teenagers at Columbine High School in Colorado is taken to be all too horridly typical; in the meantime, the Boy Scouts of America, an irreplaceable institution that has always known how to channel the healthy impulses of male aggression, and to inspire male idealism, is derided as irrelevant, "patriarchal," and bigoted.

What you get in the end is that eleven-year-old schoolboy, dutifully repeating his mantra: "We learned that you should always find a peaceful way to solve your problems because you should never be violent."

"YOU SHOULD NEVER BE VIOLENT." In this world, a world in which, to the best of my knowledge, the lion has yet to lie down with the lamb, teaching children this lesson does an unforgivable injury both to them and to the adult community of which they are about to become a part. It renders them vulnerable to abuse and injury, and leaves them without moral or intellectual recourse when abuse and injury are inflicted upon them. If no distinction is made among kinds of "peace," children are deprived of the tools they require to distinguish a just from an unjust peace, peace with honor from the peace of the grave. They are robbed of the oldest and most necessary wisdom of the race, which is that some things are worth fighting and dying for.

Are we to tell our children that, because "you should always find a peaceful way to solve your problems," the brave men who fought in the Revolutionary War, the Civil War, the two World Wars, and every other conflict in our history were acting immorally? That way lies a generation prepared only for accommodation, appeasement, and surrender. If, heaven forbid, they should ever be faced in their turn with the need to respond to aggression and evil, better by far for them to have learned, understood, and taken to heart the words of John Stuart Mill:

> War is an ugly thing, but not the ugliest of things. The decayed and degraded state of a moral and patriotic feeling which thinks that nothing is worth war is much worse. A man who has nothing for which he is willing to fight—nothing he cares about more than his own safety—is a miserable creature who has no chance of being free, unless made and kept so by the exertions of better men than himself.

What term shall we reserve for those who in the current instance have preached to us that, given who we are, and what we have done in the world, nothing of ours is worth fighting for? "Much of what is passing for pacifism," wrote the characteristically blunt columnist Michael Kelly, "is not pacifism at all but only the latest manifestation of a well-known pre-existing condition." That condition, that plague, is anti-Americanism.

CHAPTER 2

SEE NO EVIL

EVILDOERS—THAT WAS THE VERY old-fashioned, almost biblical-sounding word invoked by President Bush to describe the September 11 mass murderers. "The evil ones" was another phrase he used in those early days and weeks to describe the men who had slammed airliners into the World Trade Center and the Pentagon and hijacked the flight that was brought down in Pennsylvania by its heroic passengers. In going after the perpetrators of these despicable acts, and others like them, we were, the president said, embarking on an "effort to stamp out evil where we find it."

Not since Ronald Reagan termed the Soviet Union an "evil empire," and thereby called down upon his own head the indignation of supposedly sophisticated people everywhere, had an American president permitted himself such plain talk about our country's enemies. Enemies? In the administration of Bill Clinton, even that indelicate word had been pretty much edited out of our official rhetoric. Regimes known to be engaged in terrorism, Secretary of State Madeleine Albright declared in the year 2000, would henceforth be referred to not as rogue states but, solicitously, as states of concern.

It took George W. Bush, a "cowboy" president like Ronald Reagan, to revive the language of good and evil. Like Reagan before him, the president did so with precision and justification. For the aggression that was committed against us on September 11 had itself been undertaken not for any of the "traditional" aims of warfare—not to settle some quarrel over a disputed border, or to protest some element of policy, or to pursue some specific geopolitical advantage. No, when the advance troops of al-Qaeda set out to incinerate our innocent civilians it was not any of our deeds but the very legitimacy of our existence that for them lay under question—and it was our existence that they were seeking quite consciously, quite explicitly, to cancel. The war we were being invited to join was a war over ultimate and uncompromisable purposes, a war to the finish. Like World War II, like our war with Soviet communism, this is a war about good and evil.

Of course, the fact that President Bush was right in his choice of words did not insulate him, any more than it insulated President Reagan, from a response of disbelief and horror among the sophisticated. His statement, meant to place our response on an unassailable moral footing, had also drawn a line in the cultural sand, and our resident custodians of what is culturally permissible were quick to step up to it. Such "apocalyptic rhetoric," shot back one renowned professor of history, was scarcely less frightening than the acts of the terrorists themselves. "When Bush vilifies bin Laden," said another professor, "he's presenting a mirror image of bin Laden's rhetoric. It's name-calling." The word *evil*, instructed still another whose views we shall be inspecting, told us nothing true about the world, and certainly nothing rational. Even the word *terrorist*, according to the head of Reuters, a worldwide news agency, lacked objective meaning: "We all know that one man's terrorist is another man's freedom fighter."

"We all know." The last time I looked, there was a crystal-clear distinction between a terrorist and a freedom fighter, and it had to do with the morality of means: A freedom fighter does not massacre innocent civilians in pursuit of his ends. As for the grotesque idea that bin Laden was fighting for "freedom," try telling that to the people of Afghanistan, then groaning under the heel of his friends the Taliban. So, no, we didn't all know.

There is a formal name for the view that what is true for me is not necessarily true for you—and that name is relativism. Extended outward, from the meanings of words to the values of a whole society, it implies that we have no basis for judging other peoples and other cultures, and certainly no basis for declaring some better than others, let alone "good" or "evil." (The quotation marks themselves are intended to signify skepticism about the objective reality of these concepts.)

In one form or another, an easy-going relativism, both moral and cultural, is our common wisdom today. But things did not used to be that way. It used to be the case that a child in this country was brought up to revere its institutions and values, to identify with its customs and traditions, to take pride in its extraordinary achievements, to venerate its national symbols. What was taught along these lines in the home was reinforced in the community and the schools; what may have been wanting in the home was supplied by the community and the schools, and reinforced by public authority. The superior goodness of the American way of life, of American culture in the broad sense, was the spoken and the unspoken message of this ongoing instruction in citizenship. If the message was sometimes overdone, or sometimes sugarcoated, it was a message backed by the record of history and by the evidence of even a child's senses.

This was the common experience, the common wisdom: In the long saga of misery and inhumanity that is history (as law professor Lino Graglia has put it), the American achievement is high and unique. True, even in the past, a few of our more "advanced" mentalities, whose attitudes toward their native land had been shaped by travel abroad or association with intellectually alienated circles, claimed to see through our self-promoting cant to the less pleasant realities beneath. Some of them based their critique on a disdain for the "thinness" or the "materialism" or the "boosterism" of American

life—its "Babbittry," to use Sinclair Lewis's term—and a par-
tiality for the refinements of older, preferably European, civilizations. (This is an old theme in our literature.) Then there
were those whose taste ran to the revolutionary and the
utopian; or the still smaller number who liked to think they
had pierced the veneer of civilization altogether and for
whom nothing would really do but the authenticity of the
primitive, unspoiled by the hypocrisies of polite society or
the crassness of commercial arrangements.

But all that is now gone. Today the pyramid is inverted,
the shoe is on the other foot. Whatever may or may not be
instilled at home, little schoolchildren in our country are
routinely taught to believe that America represents but one
of many cultures and in principle deserves no automatic
preference, that there is no such thing as a better or worse
society, that cultural values different from our own need to
be understood and accepted in a spirit of sympathetic tolerance, and that, all things considered, we ourselves have at
least as much to answer for as to be proud of.

Today, it takes a considerably more nuanced mentality
to see all this—the common educational wisdom of our
own time—for the cant that *it* is, and to arrive at the reasoned conclusion that ours is, in truth, a good system, a
superior way of life, a beacon and an emblem for others.
Actually saying so can get you into trouble. It can be a risky
business for a politician, and positively foolhardy for a professor or intellectual. Even ordinary citizens have been
forced to think twice before daring to venture an opinion
on the subject.

DO YOU DOUBT IT? HERE ARE two stories from the aftermath of September 11 about the distortions introduced into everyday life by the relativist cant of our day.

In late October 2001, a school district official in St. Petersburg, Florida, intervened to put a stop to a high school physics project that was scheduled to take place on the local football field. It seems that a teacher had paid out of her own pocket for a gigantic poster bearing the likeness of Osama bin Laden. She intended to lay this poster on the football field and then, in order to demonstrate the laws of velocity and pressure, take a supply of eggs up in a thirty-five-foot-high cherry picker and drop them one by one onto a bull's-eye that had been thoughtfully painted on the terrorist's face. Her students had been working enthusiastically on this experiment for weeks, designing ways (according to a story in the *St. Petersburg Times*) "to package and protect [the eggs] from cracking during the drop."

The name of the project was The Yolk's on Osama. A little juvenile, maybe; but fair enough. So why the cancellation? "I thought perhaps, especially with our emphasis on multicultural issues . . . that it would not be a good thing to do," said the district spokesman, who then invoked the district's concerns about showing sensitivity to diverse groups and points of view, about broadcasting a "message of tolerance," and about not making value judgments.

In the end the school officials reconsidered, and the drop took place. If any Muslims were offended by the indignity to the person of Osama bin Laden, they did not make their objections known. Indeed, the entire episode was rich with unintended humor. There was the original conception of the project itself, reminiscent more of a booth at a county fair than of wartime or (for that matter) of a physics classroom. There was the prissy reaction of officialdom, with its worries about diversity and multiculturalism and political correctness. And then there was the denouement, as everyone involved seemed to sigh, "Oh, well," and give up.

I don't know what the local parents were thinking, but the absurd nature of the dispute was hardly lost on the students involved. "I understand [the planned drop] may be politically incorrect," remarked one seventeen-year-old senior, "but what [Osama] did to America was pretty politically incorrect," too.

This was a formulation it would be hard to improve upon.

MADISON, THE HOME OF THE University of Wisconsin, has a long history of leftist politics, is a very active front in the culture wars between liberals and conservatives, and was once rather wickedly characterized by a Republican governor of Wisconsin as "fifty-two square miles surrounded by reality." The quality of moral exemption implied by that crack was certainly on display in a fracas that erupted in early October over a new state law mandating that schoolchildren stand and recite the Pledge of Allegiance or sing "The Star-Spangled Banner" every day.

The law predated September 11 by a few days, but it became radioactive thereafter, as the long-suppressed patriotic feelings of ordinary Americans burst unmistakably into view. No doubt alarmed by this rising emotional tide, so alien to the official culture of the place, the school board of Madison met in early October, heard the arguments of a handful of agitated parents, and voted by three to two to override state law and ban the mandatory recitation of the Pledge. This decision was based on the most correct of politically correct grounds: protecting dissent (although the law expressly stipulated that no child could be compelled to recite the Pledge or sing the national anthem, surely, the board and some parents reasoned, an element of coercion was involved); protecting nonbelievers (who might be offended by the phrase in the Pledge, "one nation under God"); and protecting pacifists and others opposed to "militarism" (who would have difficulty with the national anthem's "bombs bursting in air").

As for the two "no" votes, they were cast only because the motion to override didn't go far *enough*—while banning the Pledge outright, it permitted schools to offer an instrumental version of the national anthem.

Unfortunately, there remained one group whose sensibilities the board had insufficently considered—namely, those Americans who had no trouble expressing their love of country in public, and no problem with their children's doing the same. Even in Madison, it turns out, such parents formed a majority, and their reaction was immediate and overwhelming. The uproar over the board's decision forced it to call another meeting, a seven-and-a-half-hour affair attended by some twelve hundred angry citizens; when that meeting began, *The New Yorker* later reported, "a large contingent of the audience rose, . . . shouted the Pledge of Allegiance, and followed that with fist-waving chants of 'U.S.A.! U.S.A.! U.S.A.!'"

By the end, the board had turned tail and voted to rescind its ban. But still it waffled, both on the merits of its original action and on the sincerity of its change of heart. Now it left to individual principals the task of deciding between offering the Pledge or "The Star-Spangled Banner," stressed that participation was voluntary, and stipulated that "those who wish to participate *may* stand" (emphasis added).

Even as the dust settled, the principal of a Madison middle school made it clear that, so far as he was concerned, not democracy but intolerance had won the day. He congratulated the board for having allowed the citizens of Madison "to exercise their First Amendment rights and express their convictions and beliefs"—a privilege for which, one supposes, those same citizens should have bowed down in gratitude— and closed with the plea that all of "us" (in which category he did not appear to include himself) learn from this episode to be "more tolerant and less judgmental."

HOW MUCH MORE TOLERANT are we expected to be? Both of these anecdotes end, thankfully, with a victory for common sense and sound majority values, but both took place after, not before, September 11 and in the face of a galvanized citizenry. That they should have happened at all testifies to the

power of the religion of nonjudgmentalism that has permeated our culture, encouraging a paralysis of the moral faculty
and leading, in the case of those school boards, to a new
tyranny of the minority.

We have already seen this paralysis at work in the neu-
tral, and morally illiterate, pronouncement of the Reuters
executive defending his news agency's decision not to call
bin Laden a terrorist. The same posture was taken to even
greater extremes by the Society of Professional Journalists, a
national organization ostensibly devoted to "the perpetua-
tion of a free press." In early October, it issued a series of
guidelines under the rubric "Countering Racial, Ethnic, and
Religious Profiling." In it, the nation's journalists, who had
not exactly been outdoing each other in displays of
unchecked jingoism, were catechized afresh in the canons of
sensitivity when it came to the values of others. Thus, they
were adjured never to write about the Muslim terrorists of
September 11 alone, as if they were a category unto themselves, but always to include a reference to "white suprema-
cists, radical antiabortionists, and other groups with a history
of such activity"; to use spellings "preferred by the American
Muslim Council" when citing Islamic names or sources; and,
most breathtaking of all, to "ask men and women from within targeted [i.e., Arab or Muslim] communities to review
your coverage and make suggestions."

Leave aside the telltale omission of the Weathermen, the
Black Panthers, or any other gang with leftist rather than
rightist credentials from the society's list of "groups with a
history of [terrorist] activity." Leave aside the fastidious dis-

tancing even from conventional Western *spellings* of Muslim terms lest they smack of ethnocentrism. Leave aside the appeal to the American Muslim Council, a political organization with a documented history of support for terrorist groups, as if it were some objective arbiter of orthographic purity. Leave aside all that. What must be unprecedented is the spectacle of professional journalists actively soliciting supervision, even censorship, from the very objects of their journalistic investigations.

"Imagine the outcry," wrote Stephen Hayes in reporting on the society's guidelines in the *Weekly Standard*, "if a newspaper editor permitted a Catholic priest to revise—before publication—a reporter's story about a pro-life rally. Or if a columnist called in a tobacco executive to edit an article about the hazards of smoking."

Yes, imagine; and then imagine this open transgression of professional ethics being codified, industry-wide. And then imagine all this being done under the name of perpetuating a "free press." The nonjudgmentalism with which some of us have allowed ourselves to become infected, and which we wear as a badge of tolerance, functions as an excuse for gross moral irresponsibility. Pretending to raise us above the "common" view, it robs us of the ability to recognize and call things by their proper names.

"Of course I dislike the Nazis," a college professor recalled a student saying to him a couple of years ago, "but who is to say they are morally wrong?" Commentators on popular culture used to help us "understand" the lessons contained in rap songs celebrating rape and the killing of policemen; multiculturalists would tell us we had no right to judge cultures practicing the genital mutilation of females; Hollywood producers would defend to my face the glorification of gratuitous violence in their movies as merely reflecting the tastes of society at large, as if they themselves had no responsibility for forming those tastes or for helping to legitimate them in their shameless pursuit of adolescent entertainment. Under the aegis of nonjudgmentalism, some Americans have ended up tolerating, protecting, or apologizing for evil—like those rap songs or those movies, or those barbarous sexual customs.

But even that is not the whole of it. Subtly or crudely, nonjudgmentalism often serves as a mask for what can only be called judgmentalism of another and much worse kind. Summoning us to some all-embracing indulgence of the views of others, however wrong or evil, it encourages us, subtly or crudely, to deprecate the good when it happens to be ours—our own values, our own instincts, our own convictions, our own civilization. To put it another way, the refusal to distinguish good from evil is often joined with the doctrine that one society—namely the United States, or the West—is evil, or at the very least that it is to be presumed evil until proved otherwise.

That is how I have learned to interpret the studied neutrality of the principle enunciated by that Reuters executive. To him, the very fact that we had chosen to call bin Laden a terrorist was enough to suggest that he should be given the benefit of the doubt—for who were we, of all people, to judge? (Coming from the world of journalism, where it is

utterly commonplace for reporters and anchormen to edito-
rialize in support of left-wing causes, this taking refuge in the
cloak of impartiality was particularly repugnant.) This also
seemed to me a key to the cluck-clucking of philosophers
and theologians who, we were informed after September 11,
were worried "that as the President casts the fight against
global terrorism as a crusade of good against evil, Americans
[would] come to feel not only morally alive, but morally
superior." And it was the message behind the storm of criticism that erupted when, breaking the nonjudgmentalist
taboo, Prime Minister Silvio Berlusconi of Italy averred in
late September that the democratic West was not just as good
as but actually "superior" to the civilization of Islam.

Coming as it did on the heels of President Bush's use of
words like "evil," Berlusconi's bold affirmation was almost
too much for the guardians of political correctness to bear.
"Simply unacceptable," harrumphed the *Washington Post*,
proposing that both the prime minister and his "deeply dangerous rantings" go back to the "different century" whence
they emerged. "Unacceptable," agreed the Belgian foreign
minister; others volunteered "absurd," "scandalous," "outra-
geous," "disgusting," and "Neanderthal." In Italy itself, the
reaction on the part of the intellectuals and the organized
Left was so violent that within days the prime minister was
forced to backtrack and apologize.

He was right in the first place, and should have stuck to
his guns. To see why, let us pause to consider the arguments
of relativists and nonjudgmentalists in defense of their own
viewpoint.

AS IT HAPPENS, ABOUT the same time as the egg-drop fiasco at the Florida high school, one distinguished relativist was mounting just such a defense in an op-ed piece in the *New York Times*. This was Professor Stanley Fish, a highly regarded scholar of English literature, dean of the college of liberal arts and sciences at the University of Illinois at Chicago, and pre-eminent spokesman of the "postmodernist" philosophy that has percolated downward and outward into our schools and into so many other precincts of our culture.

At the very moment the country was gearing up for war, amid signs of a new national unity and resolve, Fish urgently stepped in to tell us we were being . . . unsophisticated. We were being especially unsophisticated, he said, in clinging to our "mantra" that the mass murderers of September 11 were evil men. Instead, they needed to be seen as individuals "with a full roster of grievances, goals, and strategies," bearers of a rationality that just happened to differ from ours and that we happened to "reject." It was all right for us to reject it, Fish conceded: We were entitled to condemn bin Laden, and to defend our view. But what we learn from postmodernism is that "there can be no independent standard for determining which of many rival interpretations of an event is the true one." Therefore, we should abandon any "hope of justifying our response to the attacks in universal terms that would be persuasive to everyone, including our enemies."

Considered on its own terms, the position Fish articulated was almost embarrassingly weak, no different from the relativism rejected by Plato and almost every serious thinker in the centuries since then. (I choose my words carefully:

Fish himself, naturally, said that relativism was "simply another name for serious thought.") As more than one letter writer to the *Times* pointed out in the days after Fish's piece appeared, his argument refuted itself.

Take his central statement that there can be "no independent standard for determining which of many rival interpretations of an event is the true one." That is Fish's sense of things: True and final knowledge of anything is impossible. Yet he would enshrine this statement itself *as* an independent standard—the only independent standard. In a world of uncertain and relative truths, however, why should we make an exception for this one? Besides, if we did make an exception for it, as Fish urged, would we not be saying that there really *is* an independent standard for determining "which of many rival interpretations of an event is the true one," or at least the truer one? Such games of logic can go on forever, revealing little beyond the bankruptcy of the premises on which they are based.

In his article, Fish repudiated "false universals"—abstract ideas like justice and truth—as an aid to thinking about the attacks or justifying our response to them. Such ideas, he wrote, are not "persuasive to everyone," and they are not even effective, "because our adversaries lay claim to the same language" of truth and justice that we do. Better, Fish suggested, to put yourself "in your adversary's shoes, not in order to wear them as your own but in order to have some understanding (far short of approval) why someone else might want to wear them."

True, putting oneself in another's shoes does facilitate an

understanding of that person's point of view. But so what? In the first place, such an effort will always be partial at best. We can never fully exchange our moral and intellectual universe for another, for in so doing we would have to abandon our own beliefs and values. But in the second place, understanding others hardly requires us to sacrifice the appeal to abstract standards of justice. Nor does it throw us back only on our particular and oh-so-relative vision.

Of course terrorists often claim to be fighting for "justice" and to have "truth" on their side. Again, so what? We need hardly claim that our justice is perfect in order to claim, and to show, that theirs is criminal and vile. We need hardly claim that ours is the only truth in order to claim, and to show, that theirs is false and diabolical.

Is the deliberate murder of innocent civilians the same thing, morally, as the deliberate *not*-killing of innocent civilians? Is a crying baby the same thing as a ringing telephone? That is the specious sort of question we are dealing with here, and *everybody* knows the answer. To pretend otherwise is not sophisticated, it is sophistry.

If there is no independent standard, how do we know anything at all about reality? How do we know something as simple as that Stanley Fish and not, say, Toby Whale is the dean of the college of liberal arts and sciences at the University of Illinois at Chicago? And how, absent an independent standard (in this case a signed contract), would Professor Fish go about proving his title to his position in the face of Professor Whale's "rival interpretation"?

Consider a less frivolous example. Charles Manson, who

among other bloody deeds ordered the murder of the actress Sharon Tate and others one horrific evening in 1969, also fancies himself a philosopher. "I feel no man can represent another man," Manson has reasoned, "because each man is different and has his own world, his own kingdom, his own reality. It is impossible to communicate one reality through another, and into another, reality." Stanley Fish himself could hardly have put it better. Do we, then, have no independent and objective standard for determining why Professor Fish should be allowed to teach at a prestigious institution of higher learning while Charles Manson should languish in prison just because he followed a doctrine he shares with Professor Fish to its logical conclusion—the conclusion that, since everything is relative, everything can be justified and all is permitted?

One cannot help wondering what motivated this postmodernist intellectual, in the midst of war, to bring us the news that the moral convictions underlying our acts of self-defense lacked universal weight and validity. Actually, he gives us a hint as to his motive. A "growing number of commentators," Fish writes, had started to suggest after September 11 "that the ideas foisted upon us by postmodern intellectuals"—that's him—"have weakened the country's resolve." In other words, Professor Fish had perceived a threat to his *own* authority as a moral arbiter, and he was hustling to protect his turf.

National unity, moral purpose, enlightened patriotism— these are indeed threats to the "hegemony" of the Stanley Fish way of viewing the world. If they were to prevail they might end by discrediting not just him but a whole cluster of "ideas foisted upon us by postmodern intellectuals." That, to my mind, is reason enough to hope they do prevail, for until September 11 it seemed more than possible that, thanks to those ideas, the country's resolve had in truth been weakened. After September 11, all of a sudden, grounds emerged for thinking otherwise; we as a country were beginning to suit up, we were beginning to put ourselves in our *own* shoes. No wonder Stanley Fish was alarmed.

THERE IS, FOR MANY AMERICANS, a seductive quality to the relativist style of thought. I freely admit it, and I think I understand why. The doctrine that just as all men are created equal, so every idea, every opinion, every point of view is as good as every other sounds very democratic and broadminded—in short, very American. In a country that worships equality, nobody wants to seem intolerant, nobody wants to seem close-minded. For all the reasons I've cited above, however, radical open-mindedness is also a trap.

Where does an admirable respect for the views of others end and a truly crippling self-hatred begin? And at what point do we call a halt to this mindless process, lest we become complicit in our own undoing? The British journalist Robert Fisk, a sympathizer of radical anti-Western and anti-Israel causes, was covering the Afghan war when he was set upon by a violent mob of refugees who punched and kicked him and smashed his head and face with stones. "I couldn't blame them," he wrote afterward, as if they were so many little children who knew not what they did. "If I were the Afghan refugees, . . . I would have done just the same to Robert Fisk. Or any other Westerner I could find." Projecting this attitude onto society at large is the road to suicide.

It may help to remember that our Founding Fathers were themselves very open and attentive to the views of others around the world. The Declaration of Independence speaks of a "decent respect to the opinions of mankind." But the Founders did not mean by this that everyone else's opinions were as valid as theirs. That would have been to *abdicate* serious thought. Their respect for the views of others was, rather, a spur to argument, a reason to declare and defend their own "causes" before the tribunal of world opinion, and to show why those causes were universally true.

This brings me back to the "unacceptable" remarks of Prime Minister Berlusconi about the superiority of Western democratic civilization, remarks that I happen to think were both acceptable and, with certain qualifications, correct. One of Berlusconi's sins, according to his critics, was to have lumped all Islamic countries into a single "civilization" that showed no respect for human rights or religious freedom and that lagged "1,400 years behind" the West. Another was to have chauvinistically glossed over the West's own record of malfeasance and aggression. And a third was to have asserted that the West "is bound to occidentalize and conquer new people," which suggested fresh aggressions in store.

I would agree with this much: The prime minister spoke undiplomatically, and his use of words like "conquer" was unfortunate. At the same time, however, it should have been clear to anyone who bothered to read his words that what he had in mind by "conquest" was the inexorable march of Western ideas of freedom and prosperity in the modern period, ideas which, as he noted specifically and correctly, have in fact already conquered the communist world *and* "part of the Islamic world," in both cases liberating people from the yoke of state power and economic stagnation. Moreover, though Berlusconi misspoke in denying the existence of "respect for human rights and religion" anywhere in Islamic countries, it is inarguable that such respect is far and away the exception, not the rule. Finally, although the West has much to apologize for, it is the habit of apology, rather than the habit of aggression, that now seems to be a Western specialty of the house. If it weren't, Berlusconi's words would have struck his auditors as completely unexceptionable, and he himself would not have had to end up, like the true Westerner he is, apologizing for them.

The irony is that the qualities of Western culture that Berlusconi picked especially to praise are the same qualities urged upon us by the relativists and diversity-mongers. Why is our civilization superior? he asked. Because it is "a system that has guaranteed well-being, respect for human rights and . . . respect for religious and political rights, a system that *has as its values understandings of diversity and tolerance.*" Can anyone deny that these are indeed our regnant values, or that most Islamic countries do not share them? Apparently it is all right for Stanley Fish to wish them upon us, but for a Western head of state to celebrate and lay claim to them, and to breathe the word *superior*, is a scandal.

What is truth on one side of the Pyrenees is error on the other, exclaimed the French philosopher Blaise Pascal in the seventeenth century. Pascal had discovered something that would have once seemed inconceivable but that today, four centuries later, we take utterly for granted: namely, the idea that people perceive reality in particular ways and that moral judgments differ, not only among individuals but among cultures. The sociologist Peter L. Berger has usefully referred to this as the idea of cultural relativity. But relativity, Berger reminds us, is not at all the same thing as relativism. Although we cannot assume that our values are universally shared, we need not conclude either that our deepest values lack universal validity or that *no* values are universally shared.

The liberation of Kabul suggests that cultures and values are perhaps not so different after all. Anyone who saw the pictures of people suddenly free to speak, dress, learn, work, and worship as they saw fit would be hard-pressed to deny that there is a universal human longing for freedom. By the same token, simple honesty should compel us to acknowledge—and to honor—the particular cultural tradition in which that universal human longing has been most fully nourished. "We do not understand the ideals of other cultures better by misunderstanding our own," wrote the political philosopher A. E. Murphy many decades ago, "or adequately enrich an intercultural synthesis by offering to it anything less than the best we have. That best is the theory and practice of intellectual, moral, and political freedom, in a form and at a level which neither medieval, Mexican, Manchu, nor Muscovite culture has so far equaled."

The comparative study of cultures and civilizations is a quintessential product of Western curiosity. It *ought* to fill us with complex but securely founded confidence in our own culture and civilization—in its particular values, and in its universal values. If anything can be said to be a settled question in history, it is that, wherever those values have taken root, they have brought economic well-being and civil felicity in measure undreamed of. But the terrible effect of contemporary relativism—a debased and decadent product of that same admirable Western impulse of curiosity—is that, instead of imbuing us with confidence, it fills us with self-doubt and debilitates us instead. For many of its proponents, that is what it is *designed* to do. In that respect it is not relativism at all but a new and exceptionally pernicious form of absolutism.

We are under attack, and have been for some time. Not only our values are under attack, but our very existence is under attack. If, having understood the beliefs and the ideas of those attacking us, we are still not prepared to condemn them but instead find ourselves either accommodating them or throwing over them the mantle of our protection, then we might as well give up.

THE GREAT QUESTION IS TO WHAT degree any of this will change in the new era. I wish I knew the answer. I have already pointed to some signs of hopeful change, and I'll shortly point to others. But I have also pointed to signs of doleful continuity. The true believers in the religion of nonjudgmentalism are serious people, and they are not going away. In the field of education, that all-important crucible of the future, the signs so far are mixed.

Some of our institutions of higher learning are so permeated with the relativist ethos that they have finally begun to make themselves into something of a laughingstock. Even the endlessly indulgent *New York Times* could not forbear using the word *psychobabble* in reporting on some of the more ludicrous courses inaugurated after September 11 at the University of California at Los Angeles; these courses ranged from "explorations" of "America's own record of imperialistic adventurism" to examinations of "the motivations, complex behaviors, and world views of radical women" to "sharing" the instructor's own "sense of crisis about my teaching mission" in the wake of the attacks. At a moment when a *serious* respect to the decent opinions of mankind should have dictated courses in Islam, or in the history of the Crusades, or in the theory of just war, one of the country's great universities was lending itself as usual to the politics of the self and to the sort of anti-Americanism that is the inevitable consequence of relativist thinking.

Encouragingly, some undergraduates themselves are beginning to recognize, and to say outright, that they have been "had" by the relativist agenda, that it is a kind of educational scam. After a visit to Yale in late October, the commentator David Brooks had this to report:

> If I had to summarize the frustration that some of the students expressed, I would say this: On campus they found themselves wrapped in a haze of rela-

tivism. There were words and jargon and ideas everywhere, but nothing solid that would enable a person to climb from one idea to the next. These students were trying to form judgments, yet were blocked by the accumulated habits of nonjudgmentalism.

As if to put flesh on Brooks's observation, one Yale student published a column in *Newsweek* in December lamenting her generation's inability to make moral judgments. In high school, she wrote, she and her classmates had gained "emotional and psychological sophistication"—that word again!—from learning not to impose their standards on other cultures, as for instance in the matter of female genital mutilation. But now, at Yale, some of her professors and fellow students were following the same logic and refusing to acknowledge that hijacking planes and deliberately killing thousands of civilians in a terrorist act was objectively bad. As pathetic as was this young woman's idea of "emotional and psychological sophistication," her anguish at her miseducation was no less palpable, the first step on a journey out from naïveté into a truly sophisticated morality.

On some university campuses, then, there are signals of distress, and therefore of hope. Those signals are still quite faint, and they are being raised for the most part by students rather than faculty; the latter tend either to be silent or, when vocal, they tend to be (as the saying went in the 1960s) part of the problem rather than part of the solution. At Yale itself, in a faculty forum on the war, one eminent professor of history led the assembled students in an exercise of which both Stanley Fish and Robert Fisk would have heartily approved:

trying to imagine how resentful *they* would feel if they were in the position of the people "represented" by the al-Qaeda terrorists. At Williams College, my own alma mater, a Pledge of Allegiance on the Sunday after September 11 drew two hundred students, numerous maintenance and cafeteria workers, the college president . . . and exactly one professor.

All honor to those students, and to many more like them. Like the parents in Madison, Wisconsin, or like the teenage girl who defied a lifetime of her mother's leftist preachings and insisted on hanging an American flag out the window of their New York apartment, they *prove* that we as a nation do have a moral core—and they are showing it. But in university faculties, things remain largely otherwise; there, the relativist ethos of the cultural Left, despite temporary disarray, and despite the disfavor into which it has fallen in the eyes of many students, remains well entrenched. More disturbing still, that same ethos has spread, dominating the sensibility of educators on the primary- and secondary-school level.

The process is simple. It is "from their own university experiences," writes the noted education expert Chester E. Finn Jr., "that future teachers learn what to value and what not to value in education and in life." The trickle-down effect, a much-debated concept in economics, is indisputable in culture and in education: Attitudes picked up in college inevitably make their way into the fifth-grade classroom, where pupils are even less disposed than undergraduates at Yale to query their teachers' assumptions. Those assumptions, reports Finn, "forged during the Vietnam conflict and then tempered by the postmodern doctrines of multiculturalism and diversity, have overwhelmingly shaped the pedagogical and curricular guidance for elementary- and secondary-school teachers that has poured forth" since September 11.

What is the nature of that "guidance"? Leading the charge have been a number of prominent educators who rushed forward after the terrorist attack to spout relativist claptrap (the National Association of School Psychologists advised parents and teachers to refrain from suggesting that "any group is responsible"), to lecture teachers about the risks of American "militarism," or to remind us all, in the words of Bob Chase, the head of the National Education Association, the nation's largest teachers' organization, of *our* "often shameful history of vilifying our own citizens" and "cultivat[ing] prejudice and hatred on our own soil."

To grasp the magnitude of the problem as it "filters down" from adults to children, listen to one twelve-year-old girl struggling to reconcile for a *Washington Post* interviewer the disparity between what she had learned at home about the events of September 11 and what she was being taught in school: "They wanted to bomb our symbols. That's what my mom said. Because we're bossy. That's what my teacher said." And here is a deputy schools chancellor in New York City: "Those people who said we don't need multiculturalism, . . . a pox on them. I think they've learned their lesson [after September 11]. We have to do more to teach habits of tolerance, knowledge, and awareness of other cultures."

I have tried in this chapter to show the fraudulence of

this way of thinking. It goes very deep; and so does the long-term investment in it, on the part not just of our educators but of our media and still other institutions. (Remember "one man's terrorist is another man's freedom fighter"?) Rooting it out, replacing it with healthier growths, will be the work of generations.

Our war, if it proves as successful over the long haul as it has so stunningly proved in the short term, will certainly help accelerate the great relearning, restoring moral clarity and deepening our confidence in our civilization and what it stands for. But the real work will have to be done at home, and preeminently in our schools and universities. We will know a corner has been turned when a future head of the National Education Association begins speaking of our national history not as "often shameful" but as morally purposive, self-correcting, and glorious. We will know that things have really changed when good and evil are once again called by their proper names, and when Americans, children and grown-ups alike, are no longer taught to be confused about which is which.

CHAPTER 3

A WAR AGAINST ISLAM?

THE SCENES WERE VISIBLE FOR only brief moments on our television screens, but they were stunning and unmistakable: crowds of ordinary Muslims around the world, joyously celebrating the fall of the twin towers, the extinction of American lives, and the humiliation of the "Great Satan." Other images were quickly mobilized to replace those images of exultation, as various Muslim and Arab spokesmen came forth to express their solemn sympathy for the victims. But few if any of these spokesmen seemed ready to connect the bloody deeds of September 11 with the religion or culture of Islam, and some of them proceeded to reverse field and invoke a different culprit altogether. This culprit was a heretofore unknown phenomenon with the cumbersome name of Islamophobia.

Had not war been declared by Osama bin Laden against the infidel West? No, we were told, it was the other way around: The West was at war with Islam, and our response to September 11 had only proved it. The events of that day, declared one spokesman, had provided an excuse for yet another eruption of "anti-Islamic vitriol" in the long "smear campaign" against all Muslims. The root problem, said one writer, lay in the "malevolence" of Westerners, who "routinely demonized" Islamic faith and practice and were now calling into question the loyalties of millions of Muslim-American citizens. Taking a somewhat different tack, Edward Said wrote that the "backlash of September 11" featured "an almost palpable air of hatred directed at [Arabs and Muslims] as a whole," with "stereotypes of lustful, vengeful, violent, irrational, fanatical people" everywhere in evidence; to this literary celebrity, whose numerous books have sought to portray a centuries-long Western assault upon Islam and Arabs, the American reaction to September 11 thus seemed to suggest a mere variation on a very old theme.

Bin Laden himself had said that we were the heirs of the medieval Crusaders. Had not the U.S. president, as scores of Islamic voices rushed to chide us, confirmed bin Laden's point when he had the insouciance to refer to the war against terrorism as a "crusade"?

THE PRESIDENT'S FLEETING USAGE of the word *crusade* was unfortunate but wholly innocent; his real view of the matter was very much otherwise, and he stated that view with characteristic directness in his address to Congress not long after the mass murder of September 11. The terrorists, the president said then, represented "a fringe movement that [had] pervert[ed]" the essentially peaceful teachings of their religion, and their "form of Islamic extremism . . . has been rejected by Muslim scholars and the vast majority of Muslim clerics." During a visit to a mosque on September 17, he also went out of his way to praise the loyalty of the "millions of good Americans who practice the Muslim faith who love their country as much as I love the country."

The president was hardly alone in voicing these sentiments. "Islam is a peaceful and tolerant religion," echoed British prime minister Tony Blair, "and the acts of [the terrorists] are contrary to the teachings of the Koran." In dozens of editorials, our leading newspapers likewise hammered home the same series of points: Islam is a religion of peace, and in waging war against terrorism we were not at war with that religion; in no way did the bin Laden terrorists represent "mainstream" Islam; American Muslims had categorically condemned the terrorists, and the rest of us had to be especially careful to safeguard their rights and to see that they suffered no discrimination.

There were, of course, some who held otherwise, particularly in the heated aftermath of the attack. Franklin Graham, the son of Billy Graham and the head of the Billy Graham Evangelistic Association, proclaimed Islam "a very evil and wicked religion," and even after backing away from that statement he continued to assert his "responsibility to speak out against the terrible deeds that are committed as a result of Islamic teaching." Saxby Chambliss, a Republican congressman from Georgia, proposed that we let the local sheriff "arrest every Muslim that crosses the state line." Among Muslims themselves, both here and abroad, there were those who gave the lie to the president's characterization of them by openly proclaiming or advocating war with the West.

But in this country the weight of centrist opinion was and has remained as I have stated it: that the hijackers of September 11 had also hijacked their own religion, and that it was incumbent on us both to acknowledge that fact and to understand and appreciate what a *New York Times* article by the religious historian Karen Armstrong called "The True, Peaceful Face of Islam." This injunction was a prelude to another: that we must also come to understand, and to transcend, our own Islamophobia.

Do we suffer from such a disease? Are we engaged in a war against Islam? Even framing the questions in those terms may suggest the same sort of obsession with "us," and especially with our sins, that I have spent the last two chapters trying to expose and refute. In what follows I am guided by a different principle: that the best way to understand others is to assume that they are as fully responsible for their own behavior as we are for ours.

I WOULD BE A FOOL TO CLAIM any deep expertise in a religious civilization whose vivid and complex history stretches back almost fifteen hundred years, whose achievements during at least the first millennium of its earthly career dazzle the imagination, and whose adherents today number over a billion people in countries all over the globe. What I do know more than enough to say is that, like the other great revealed religions of Judaism and Christianity, Islam *is* a faith of history, an evolving faith, and one whose countenance shows different facets at different moments while validating itself at all times by reference to its founding documents and sacred traditions.

This is hardly to suggest that, any more than Judaism or Christianity, Islam is or can be all things to all people, or that one cannot locate and define its "essence." Rather, in speaking of it, one must be sensitive both to that very complicated essence and to the variegated circumstances, and tonalities, in which the essence has been historically expressed.

For example, much has been made lately of the concept of jihad, loosely translated as "holy war" and invoked by many contemporary Islamic spokesmen as a continuous duty incumbent on all Muslims in their relations with non-Muslims until the whole world is brought under the mantle of the Prophet Muhammad. There can be no doubt that this concept is strongly articulated in the Koran. Nor can there be any question of its cardinal importance in later Islamic thought; although rough counterparts to the idea may be discerned in other religious traditions, and a closer parallel in modern revolutionary ideologies like Leninism and Maoism, as a fully developed religious doctrine jihad is unique, and central, to Islam. Nevertheless, whole periods of history have passed—would that ours were one of them!—in which this doctrine has been sublimated into less activist forms or has slipped from a position of centrality in public discourse, in religious preaching, in political debate, or in mass consciousness.

We have been well cautioned against confusing a fringe movement, the Islamic terrorists, with the center, Islam itself, lest we commit the intellectual and moral crime of "scapegoating" a whole religion that, like other great religions, stands only for comity and peace. There is, however, a more honest and less blinkered course. Let us grant that extreme tendencies or indeed entire heresies can arise that occlude more balanced interpretations of the center. The question then becomes one of assessing how much influence such tendencies exert over the center, and whether the center fights back. At the same time, it does no good to pretend that religions and cultures are everywhere basically the same, and basically the same as the ones we happen to know. Still less helpful is it to ignore the particular nature of the center itself, in all its solidity and obduracy. A greater realism about these matters not only shows more respect for genuine human differences but, paradoxically, holds out greater hope of reconciling those differences in the end.

LOOMING HUGELY OVER THE landscape of contemporary Islam for at least the past three decades has been a militant, jihad-based ideology, profoundly hostile to religious tolerance or pluralism of any kind, that has been given the name of "Islamism." Its specific roots have been traced by some scholars to Wahhabism, a puritanical and xenophobic tendency within Islam that arose in the eighteenth century and flourishes in modern-day Saudi Arabia, where it is embraced, subsidized, enforced, and propagated by the regime. The goal of this ideology is to establish an extreme totalitarian theocracy in all Muslim lands and to extend dominion outward to the free and open societies of the West, which are the special objects of its atavistic rage—and none more so than the United States of America, also known as the Great Satan.

There is no reliable way of estimating the overall size of this tendency: The Muslim world is vast and varied and runs the gamut from the Iran of the ayatollahs to secular and largely westernized Turkey. The Islamists' hard-core and terrorist cadres are, moreover, relatively few in number. But they have drawn on vast financial resources, including from the more moderate Muslim regimes seeking to protect themselves, and they enjoy an enormous popular following almost everywhere in the Muslim world, especially in places, and these are very plentiful, where material conditions are woeful. As distinct from the hard-core activists, that sympathetic following has been authoritatively gauged in the hundreds of millions.

Among the many vanguard institutions of this ideological movement—from Osama bin Laden's worldwide al-Qaeda to Hamas and Hezbollah and Islamic Jihad in the Middle East and North Africa, to the Islamist insurgencies in Indonesia and the Philippines, to their brethren in Central Asia and Pakistan, to the regimes holding actual power in Iran, the Sudan, and (until yesterday) Afghanistan—there is no question that Islam is at war with the West and, specifically, with America. The Islamist government of Iran alone has poured forth a decades-long torrent of vituperation against the United States; is implicated, according to our State Department, in the deaths of Americans at a U.S. Air Force barracks in Saudi Arabia that was bombed in 1996; and as of the end of 2001 was sheltering seven of the twenty-two suspects on the FBI's "most wanted" list issued after September 11 for acts of terrorism directed against Americans. As for Osama bin Laden, in a 1996 religious edict he specified "no more important duty" for Muslims than removing American soldiers from the "holy places" of his native Saudi Arabia. Two years later, he called on "every Muslim who believes in God and hopes for reward to obey God's command to kill the Americans." Three years after that, having just made good on his threat in the most horrific way imaginable, he besought God to bless the terrorists who had massacred thousands of innocent Americans and "allot them a supreme place in heaven." The attacks of September 11, he concluded, had "divided the world into two camps, the camp of the faithful and the camp of infidels, may God shield us from them. . . . God is the Greatest and glory be to Islam."

In the mind of more than one Islamic radical, then, there is definitely a war between Islam and the "infidels," and it is a war mandated by religion. Nor did this war "start" on September 11, though that was when, very belatedly, we ourselves were compelled to join it. To restrict the list of hostilities to American targets alone, we must never forget the earlier bombing of the World Trade Center in 1993, or the bombing of the U.S. barracks in Saudia Arabia in 1996, or the suicide assaults on two American embassies in Africa in 1998 and on the USS *Cole* in Yemen in 2000. The earliest of these operations, as the Arabist scholar Martin Kramer wrote to no avail in 1993, should have brought home the folly of the then-fashionable distinction between "moderate" and "extreme" Muslim militants, and the absolute need to "err on the side of caution" in protecting the safety of our citizens.

Islamist radicals have targeted others besides Americans. There are violent Islamist groups in Egypt, Saudi Arabia, Yemen, Somalia, Eritrea, Djibouti, Afghanistan, Pakistan, Bosnia, Croatia, Albania, Algeria, Tunisia, Lebanon, the Philippines, Tajikistan, Azerbaijan, Chechnya, Uzbekistan, and Kashmir. In Algeria, the brutal struggle between radical Islamic forces and a repressive government has yielded 100,000 casualties over the last decade, and it is not over; in Uzbekistan, the hard-core Hizb-ut-Tahrir is implacably bent on Islamizing the secular regime. Wherever Islamists operate, their goal is to expel from Muslim areas not only Westerners and non-Muslims but also "deviant" Muslim leaders who do not subscribe to Islamist beliefs. Still, we are their main target.

I have not forgotten that openly Islamist terrorists are not the only terrorists we face. Nor have I forgotten that states like Syria and Iraq, which support and harbor and train terrorists, are secular regimes; or that Yasir Arafat, an arch-terrorist himself and the godfather of modern terrorism, is no Islamist, although he has inveterately helped himself to the rhetoric of violent jihad whenever it has suited his purposes. Finally, I have not forgotten President Bush's stricture that ours is not a war against Islam. All that being stipulated, I am also mindful of the words of the exiled Iranian author Amir Taheri:

> [I]t is impolite, not to say impolitic, to subject Islam to any criticism. Yet to claim that the attacks [of September 11] had nothing to do with Islam amounts to a whitewash. It is not only disingenuous but also a disservice to Muslims, who need to cast a critical glance at the way their faith is taught, lived, and practiced. Even worse, the refusal to subject Islam to rational analysis is a recipe for further fanaticism.

To a greater extent than we have permitted ourselves to say, this war has to do with religion.

ACCORDING TO THE EMINENT scholar Bernard Lewis, whose work I rely upon heavily here (and whose essay in *Commentary*, "The Return of Islam," a prescient analysis of things to come, appeared as long ago as 1976), the Islam of the September 11 killers "is not the true Islam. But it comes out of Islamic history and culture." This makes it incumbent upon us to familiarize ourselves with pertinent elements of that history and culture.

To begin at the beginning, does the word *Islam* mean "peace," as one of former President Clinton's speechwriters put it? In fact it means "submission"—in particular, submission to the will of Allah. Having been vouchsafed a revelation of that will, the Prophet Muhammad founded a religion inspirited by the zeal to succeed and propelled by a militant confidence in the rightness of his cause. And indeed, for the first few centuries of its existence, Islam was an almost unstoppable force—and force, including the murder or conversion of masses of people at swordpoint, was regarded as the wholly legitimate tool of its ascendancy. In rapid succession its armies conquered the lands of Persia, Egypt, and Syria, eventually sweeping across the North African littoral, spreading into Spain, and moving inexorably northward until halted by Charles Martel at the Battle of Tours in 732.

Over the next several centuries, Islam and Christianity traded victories and defeats, with Crusader thrusts into the Holy Land and elsewhere being more than offset by Muslim reconquests and the takeover of India and the eastern parts of the old Roman Empire. During this period Islam reached peaks of intellectual and economic accomplishment, anticipating, rivaling, or surpassing the achievements of Christian Europe. But decay had set in even before the decisive defeat of advancing Muslim forces at the gates of Vienna in 1683. That event heralded an even longer period of decline, colonization, and loss, highlighted in the early twentieth century by the traumatic breaking-apart of the Ottoman Empire at the hands of the Western powers after World War I.

It would be hard to overestimate the significance in Muslim consciousness of this series of defeats: When Osama bin Laden alluded cryptically in an October speech to the dissolution of the Ottoman Empire, he could count on the ominous resonance of his words among his Arabic-speaking listeners, for in the classical understanding, the world is divided into the "realm of Islam" and the "realm of the sword"—that is, the realm not yet conquered by Islam—and no piece of land, once fallen to the former, can be renounced or alienated or surrendered to the latter. (Although President Bush was sharply criticized for offending Muslim sensibilities when he used the word *crusade* to characterize our war against terrorism, it is plain that we harbor no plans to conquer anyone, or to separate anyone from his religion; the same cannot be said for Islamist radicals.) Similarly, a penalty of death awaits the convert to Islam who reverts to his previous religion.

Note, in the previous paragraph, the almost reflexive joining of political considerations (land, power) with spiritual ones (adherence to a religious faith). I have referred once already to this conjunction in discussing comparative attitudes toward pacifism among the monotheistic religions, but it bears emphasis again. In theory, if not in practice, Christianity as a faith was always distinct from Christendom as a political reality; by contrast, the "world of Islam" comprises and combines temporal and spiritual concerns. "From the lifetime of its founder," writes Bernard Lewis, "Islam *was* the state, and the identity of religion and government is indelibly stamped on the memories and awareness of the faithful."

To these faithful there is but one Muslim world or empire, artificially divided into nations and ethnic groupings but united in essential loyalty and identification. It is not a question of being an Arab or a Turk—indeed, the fragmentation of the Islamic empire into national groupings with their own

separate interests is one of the many alleged disasters perpetrated by infidels that Osama bin Laden meant to correct. It is,
rather, a question of being a Muslim *or* being an infidel.
Between the two there exists a state of permanent war.

To be sure, the nature of that war is governed by strict
codes, adumbrated in the Koran itself and refined over centuries of reflection and interpretation. Killing innocents, for
example, is forbidden, and an offensive jihad comes under
tighter limits than a defensive one in which "they [the unbe-
lievers] attack you [the faithful] first." Terrorism is ruled out,
as is suicide, suggesting in Lewis's words that "there is no
precedent and no authority in Islam" for the assaults of
September 11. On the other hand, martyrdom in the service
of Allah has always been honored among Muslims, and the
category of the martyr in Islamic thought, unlike in Christian
thought, routinely includes one "killed in battle."

An echo of all these theological distinctions turned up in
Osama bin Laden's ludicrous post–September 11 claim that,
in massacring innocent Americans, "we are only defending
ourselves. This is defensive jihad." (Ghoulishly, he went on:
"This is a simple formula that even an American child can
understand. Live and let live.") It is also no doubt why so
many bin Laden sympathizers are at pains to claim that
Islam is itself under attack by an "Islamophobic" West.

Bin Laden aside, some Islamic scholars today have
indeed elevated suicide bombers to the ranks of martyrs in a
holy war, especially if the suicide bombers happen to have
targeted innocent Israeli civilians rather than innocent
American civilians. Because "all Jewish women in the Zionist

entity" are by definition "fighters," went one religious ruling from Kuwait in late August 2001, "martyrdom attacks that aim to kill the occupiers in order to strike terror in their heart" may legitimately (though of course unintentionally) kill "women and children . . . with full right." Whether or not they actually celebrate such deeds, few are the clerics in the Muslim world willing to say outright that suicide bombings are banned by the Koran, or that (in the words of one such lonely cleric, Cairo's Sheik Muhammad Rafaat Othaman) "attacking innocent, unarmed people is forbidden" under any circumstances. As the journalist Douglas Jehl discovered at a conference of Arab intellectuals in Cairo in late November, fewer still—practically nonexistent—are those prepared to offer a "sustained and comprehensive rebuttal of the radical theology" of bin Laden and the Islamists.

For many in the Islamic orbit, in short, what defines the religious status, and the morality, of a terrorist act is whether a given target is "legitimate," and the sole criterion of legitimacy would seem to be a perceived sufficiency of grievance. The consensus appears to be that terrorism against Israel qualifies without hesitation; against America, with hesitation.

But, it will be said, this is understandable enough: People bent on a course of action naturally seek justification where they can, and besides, every religion has its contradictions and its hypocrisies. So why make a special issue of Islam's? Look, after all, at the long record of violence conducted in the name of Christianity, the religion of love and peace and forgiveness.

But that is just my point. Christianity's record is indisputably spotted. But standing in eternal reproof of that record, crying hypocrisy and betrayal, is Christianity itself, quintessentially embodied in the example and the teachings of Jesus. In the case of Islam, the charge of hypocrisy hardly applies—certainly not on the matter of religious violence. To put the issue at its starkest, there is simply no equivalent in the Koran to the New Testament's admonishment to "turn the other cheek"; conversely, there is no equivalent in the New Testament to the Koranic injunction to "kill the disbelievers wherever [you] find them."

THIS BRINGS US BACK TO THE question of whether the brand of radical Islam represented by Osama bin Laden was indeed an artificial growth that "hijacked" the classical faith. What I have been trying to suggest is that the growth is not artificial, and that the classical faith itself is not without its deeply problematic aspects, particularly when it comes to relations with non-Muslims. The superiority of Islam to other religions, the idea that force is justified in defending and spreading the faith—these teachings have been given high visibility in Wahhabism, but they are authentic teachings.

True, in classical Islamic theory, Jews and Christians are at least accorded an officially tolerated (*dhimmi*) status as adherents of monotheistic religions, but in practice the attitude toward them has always been condescending at best, ranging from the contemptuous to the murderously hostile. In schools all over the Islamic world today, even in universities, triumphalist habits of thought are inculcated as a matter of course, reflective self-criticism is unknown and unencouraged. Hatred of Jews is vicious and endemic and defining; in many Muslim countries, especially in the Middle East, Jews—not Israeli Jews but Jews, period—are not even permitted to reside. Conspiracy-thinking runs rampant; just this past fall, during the second half of Ramadan, Arab television broadcast a lavishly funded thirty-part series, *Horseman Without a Horse*, dramatizing the notorious nineteenth-century forgery known as *The Protocols of the Elders of Zion*. In faraway Malaysia, Prime Minister Mahathir Mohammad once blamed Jewish control of the media for criticism of him in the Chinese (yes, the Chinese) press.

What Islam needs and has never undergone, some have argued, is the equivalent of the eighteenth-century European Enlightenment. Christianity made its peace with modernity; in time Islam will, too. This indeed may happen; if it does, it may begin to take shape first among Muslims in the rich cultural stew of the United States, where a few voices have already asked, in the words of a recent op-ed article by a former Egyptian journalist, "What in Islam, what in the way it is practiced today, allowed bin Laden to promote his murderous message?" For many, however, radical Islam of the bin Laden type, for all its insistence on a return to a pure and austere faith, is itself (as Clive Crook noted in *National Journal*) "a modern invention," not a remnant of a past to which people blindly cling but an innovation that shrewdly picks and chooses among the offerings of modernity. So far, and despite our great initial successes in the war against terrorism, it has shown few signs of retreating.

Even some who grant that Islam itself, and not just Islamism, is implicated in the events of September 11 tend to leave out what makes it unique, or to combine its more discomfiting manifestations with other religious or political phenomena with which they happen to be familiar. Thus, the social theorist Robert Wright argued in *Slate* that violence is a strain present in all religions, and that before the industrial revolution, Christianity, too, was marked by an "intense authoritarianism." "Calvin," Wright asserts, "ruled Geneva roughly as Stalin ruled Russia." Andrew Sullivan, formerly the editor of the *New Republic*, who has shown a much keener appreciation than Wright of the theological differences between Islam and Christianity (and who has also disagreed with Wright that the key to promoting religious tolerance is economic and social development), has nevertheless contended that the problem we face is not Islam but religious fundamentalism in general, a reactionary tendency at war with "faiths of all kinds that are at peace with freedom and modernity." It is, wrote Sullivan, "almost as if there is something inherent in religious monotheism that lends itself to this kind of terrorist temptation." A secularized version of this same argument has been echoed by many other writers who speak (in the words of Michael Lind) of "a war of reason and tolerance against medieval superstition" and draw parallels between the Taliban and religious conservatives here in the United States.

Well, when it comes to superstition and authoritarianism we have a rich menu to choose from. The most vicious authoritarian regimes of the past century were not religious;

they were dogmatically atheist. Indeed, the communists hated and were terrified by revealed religion, and for good reason: In it, and especially in the teachings of Christianity and Judaism, the value of every living thing, our individual preciousness in the sight of God, is proclaimed in accents that cannot be brooked and that strike at the heart of every tyranny.

Interestingly enough, the communists seemed to have less trouble in this regard with Islam. In a striking passage in her magnificent memoir *Hope Against Hope* (1970), Nadezhda Mandelstam recalls how her husband Osip Mandelstam, perhaps the greatest Russian poet of the twentieth century and a man who paid with his life for a verse criticizing Stalin, was drawn to such outlying parts of the Soviet empire as the Crimea and Armenia, with their still-living links to ancient Greek and Christian tradition. In this, she remarks, Mandelstam differed from most Soviet and communist writers, who, when they "traveled to the borderlands—something that was very popular—chose the Muslim areas." Her husband, she comments,

> thought this preference for the Muslim world was not accidental—the people of our time were less suited by Christianity with its doctrine of free will and the inherent value of the person than by Islam with its determinism, the submerging of the individual in the army of the faithful, and the formalized design of an architecture which made man feel insignificant.

As for the temptations of power, most fundamentalist Christians in the United States do not seek power; if any-thing, they tend to abhor it, to shun its nosy intrusions into their lives. Finally, although it is true that a very few funda-mentalists have resorted to violence—the favored example is of those who murder abortion providers—such violence, as Andrew Sullivan himself writes, has not been a "significant feature" of religious fundamentalism in this country. Moreover, some of the most prominent figures condemning it have been those, like Pope John Paul II and the Catholic bishops, who likewise oppose abortion but who have repu-diated these fanatics as wholly outside the boundaries of the faith they claim to uphold. When it came to the mass mur-derers of September 11, far too many voices of true authority in Islam were silent if not approving.

Still other observers have urged us to distinguish between militant Islamists and . . . militant Islamists. Thus, law professor David Forte writes of a "difference between religious imperialism and intolerance, on one hand, and fly-ing planes into the World Trade Center and perhaps unleash-ing a plague upon our people on the other"—in other words, between intending or encouraging others to impose one's beliefs and actually doing so by killing unbelievers. No doubt there is such a difference, but it could hardly be plain-er that the intention is linked to the act. In the words of Sheikh Muhammad Hisham Kabbani, one of the very few truly moderate Muslim clerics in the United States with the courage to speak out, "Wahhabi belief provides the religious and ideological underpinnings to enable militant movements to take up arms against existing governments if they deem the need arises."

It is estimated that as many as two-thirds of the people of "moderate" Saudi Arabia, the nation regularly cited as our closest ally in the Arab world, adhere to Wahhabism, and Saudi Arabia itself is ruled by a strictly enforced brand of Islamic law; no other form of worship is legal in the kingdom. So it is no surprise that, as was reported in the *New York Times*, the average Saudi should view Osama bin Laden as "the conscience of Islam," "a hero, the ideal Muslim, the epitome of what an Arab should be." According to an influential Saudi lawyer, "What [bin Laden] says and what he does represents what most Muslims or Arabs want to say and can't. What he says we like, we agree with it."

But what about elsewhere in the Arab world, where different and presumably more moderate forms of Islam are practiced? Things, alas, are no different. "Reporters from Arab shores," writes Professor Fouad Ajami of Johns Hopkins, "tell us of affluent men and women, some with years of education in American universities behind them, celebrating the cruel deed of Muhammad Atta and his hijackers." Here is a columnist in an Egyptian newspaper: "I cannot restrain my joy. For the first time in my life, I witness with my own eyes the defeat of American arrogance, tyranny, conceit, and evil." "Allah decreed the vengeance against you," echoes a columnist in Egypt's *Afaq Arabiyah*. "Muslims see what happened as divine retribution." A Libyan confided in Joseph Lelyveld of the *New York Times* that "September 11 was the happiest day of my life," while a policeman for the Palestinian Authority called bin Laden "our messiah." Wrote a Palestinian journalist in late November: "It is virtually impossible for me, as indeed is the case for most Palestinians, Arabs, or Muslims, not to hate America."

And outside the Arab sphere, in the broader Muslim world? The visage of bin Laden was on T-shirts everywhere, and after September 11, Osama reportedly became the most popular name for baby boys. "Bin Laden is Islam" was the cry in the streets of Pakistan, where the Islamist academies, or *madrasas*, were busily indoctrinating future generations of extremists and America-haters. In northern Nigeria, Reuters reported, bin Laden had achieved "iconic status." In Indonesia, half the respondents to an on-line survey considered him a "justice fighter," only 35 percent a terrorist. In Kenya we heard: "Every Muslim is Osama bin Laden." Meanwhile, much closer to home, a survey conducted in Great Britain by the (London) *Sunday Times* found four out of ten British Muslims agreeing that bin Laden was justified in waging war against America, and an equal proportion believing that British Muslims were justified in going to fight alongside the Taliban.

And so it went. No wonder the *Washington Post* was driven to conclude that almost everywhere in the Muslim world, radical and nonradical alike, whether one consulted ordinary people or intellectuals and religious leaders, bin Laden was cheered "with almost a single voice."

IT IS NOT ACCURATE TO SAY, THEN, that Islamic extremism has been rejected by the overwhelming majority of Muslims, let alone by the overwhelming majority of Muslim clerics, around the world. It is not even accurate to say it has been unequivocally rejected by the overwhelming majority of Muslim leaders, clerical or secular, here in the United States.

For three and a half years, Mohammad Al-Gamei'a served as the chief imam of the Islamic Cultural Center in New York, a Sunni mosque on Manhattan's Upper East Side that is attended by about two hundred people daily and three thousand people each Friday, among them many members of the U.N. diplomatic community. Soon after September 11, the imam departed mysteriously for Egypt, where he gave an interview to a website affiliated with the prestigious al-Azhar University. There he blamed the terrorist attacks on "the Jews," who, he said, wanted to "impose their hegemony and colonialism on the world" and were "the only ones capable of planning such acts." Although, the imam continued, the "Zionist media" tried to cover it up, "we know" that Jews "disseminate corruption in the land" and "murdered the prophets; do you think they will stop spilling our blood? No."

If the imam sounded a bit unhinged—in the same interview he suggested that Muslims in America were afraid to go to hospitals lest they be poisoned by Jewish doctors, and that Jews were disseminators of "heresy, homosexuality, alcoholism, and drugs"—his successor at the mosque, a former translator at the United Nations named Omar Saleem Abu-Namous, was only somewhat more reassuring when interviewed in October by Rachel Donadio of the weekly *Forward*. Asked his opinion as to who was behind the attacks, the new spiritual leader conceded that Osama bin Laden was "one possibility," but then pleaded ignorance as to whether the massacre was "committed by Jews or non-Jews, because I don't have any evidence." As for the nineteen terrorists "who were supposed to have died, many of them are still living, in point of fact. . . . The reports were not precise." At that point in the interview, the imam switched to a subject with which he evidently felt more comfortable, namely, the suicide bombers who target innocent Israeli civilians and who are, he assured his interlocutor, "martyrs" fighting for a "rightful cause."

Elements of the imam's peculiar vision are regrettably shared—scandalously shared, I would say—by a number of the so-called mainstream organizations that claim to represent the interests of Muslims in America, of which the best known are the American Muslim Council (AMC), the Muslim Public Affairs Council (MPAC), and the Council on American-Islamic Relations (CAIR). It was the leaders of these organizations who stood by President Bush's side when he appeared at the Islamic Center in Washington within a week of the September 11 attacks and expressed his belief that "they feel just the same way I do . . . they love America just as much as I do." But as was subsequently reported in *Salon* in late September, CAIR was quite equivocal in its statements and would only say, "If bin Laden was behind it, we condemn him by name." CAIR and AMC have defended or supported such terrorist groups as Hamas and Hezbollah, and some of these organizations had made a habit of decrying supposed anti-Muslim bias in American prosecutions of earlier terror suspects. In 1994, the executive director of the AMC, Abdurahman Alamoudi, had addressed a cheering rally outside the White House in these words: "Hear that, Bill Clinton! We are all supporters of Hamas. I wish to add that I am also a supporter of Hezbollah. Anybody support Hezbollah here?" And the crowd cheered.

Does this mean that American Muslims in their majority subscribe to the Islamist agenda? Hardly. But it does mean that if any "hijacking" has occurred, it has been perpetrated by those who speak *in the name of* American Muslims. I have mentioned one exception, Sheikh Muhammad Hisham Kabbani: In 1998, speaking to the State Department, Kabbani estimated that extremists had taken over fully 80 percent of the mosques in this country and exercised ideological control over schools, youth groups, professional associations, and commercial enterprises. CAIR argues that no one can point to any example of a major American Muslim organization supporting Islamic extremism, but others disagree. In the summary judgment of the analyst Daniel Pipes, who has documented this deeply alarming phenomenon in copious detail, "the major Muslim organizations in the United States are in the hands of extremists"; their shared, militant outlook is "hostile to the prevailing order in the United States and advocat[es] its replacement with an Islamic one."

Schools: Just outside Washington, D.C., in one of America's most beautiful neighborhoods, is the Muslim Community School of Potomac, Maryland. To an inquiring reporter for the *Washington Post*, the principal of this school declined to blame bin Laden for the attacks, referring instead to a guidesheet that found "as much evidence pointing away" from the terrorist as "circumstantial 'evidence' pointing toward him." "Whatever is said about" bin Laden, the principal added, "I want it said about the Israeli prime minister." As went the principal, so went the students. One eighth-grader told the reporter that if forced to choose between his country and his religion, "I'd stay with being Muslim. Being an American means nothing to me."

In Brooklyn, New York, a reporter from the *New York Times* encountered similar views. "'Isn't it ironic that the interests of America are always against what Muslims want?'" asked one seventeen-year-old, who assured the reporter that "he would rather go to jail than fight in the United States Army against Muslims." Others said they "would support any leader who they decided was fighting for Islam," and all insisted they "were not convinced that Osama bin Laden—or any Muslim, for that matter—was behind the attacks on the World Trade Center and the Pentagon."

These are truly disturbing signs of alienation and disaffection, and they evoke the very stereotypes of Muslim Americans that organizations like CAIR are constantly accusing others of inventing. What is heartbreaking is how few countervoices exist within the Muslim community itself. Religious authorities like Kabbani, uncompromising scholars like Fouad Ajami, newly energized critics like Muqtedar Khan of Adrian College, Michigan—such people deserve the gratitude of us all. They need to be encouraged, and their numbers need to grow. But they have their work cut out for them. They face not only the entrenched power of the victimology-mongers at CAIR and elsewhere but the no less insidious predisposition of many

American professors, intellectuals, and journalists, who have long tended to look with benign tolerance on radical movements around the world that speak in the name of the "disenfranchised" and who locate the source of every human problem in the evil wrought by corporate America or the Pentagon.

"Being an American means nothing to me." When an eighth-grade teenager living in an affluent American suburb speaks like this, we have a problem. When I quoted this teenager's remark and expressed my opinion of it on a television program, I was told that dissent was a good thing and that I should be "more tolerant." That makes two problems.

THERE IS NO QUESTION THAT some Muslims who live among us do bear a burden. For those who are strictly pious, there are bound to be difficulties associated with fulfilling the commandments of their religion in a secular state and in a society in which the majority is Christian. Those difficulties are compounded by the strictures imposed by Islam on relations with "infidels." And then there are the special issues created by the cataclysm of September 11 and the steps that have had to be taken by the federal government to ensure the security of all our citizens, including immigration restrictions, scrutiny of young men from the Middle East, "profiling," and screenings and detentions of suspects for questioning.

Troublesome as these various obstacles may be, they are hardly crippling or insurmountable. Nor is the situation of

Muslim Americans in general a disadvantageous one—far from it. Economically, the community is an unqualified success story, and both socially and politically it has developed a significant presence on the American scene, where, as it happens, its concerns have been accommodated to a perhaps unprecedented degree in the history of ethnic groups in our country.

Over the last decade, Muslim Americans have won court case after court case over such matters as being allowed to wear beards or, for women, headcoverings at work; being excused for prayer several times a day; being provided with special locations for prayer in their workplace; and so forth and so on. American officialdom at every level has gone out of its way to express a warm welcome to Muslim immigrants, to support them in finding work and joining communities, to include Islamic symbols and Muslim representatives in major public rituals. The American media treat Islam and Muslims with extraordinary sympathy. Ordinary Americans have conducted study sessions, mounted library exhibits, invited Muslim preachers to their churches and synagogues, and in general displayed the tolerance and generous understanding of strangers for which they are well known and which have made this country a mecca (if I may be permitted) for people around the world.

As for the restrictionist measures taken after September 11, these have been vociferously debated in the American press, and the American Civil Liberties Union, which opposes them, has gone out of its way to offer its rather formidable help to anyone who feels mistreated by our government.

That some of these measures represent compromises with our sound liberal traditions, or at least with recent interpretations of those traditions, is undeniable. It is not true, however, that they represent compromises with our traditional practice in wartime, when a price has always had to be paid for ensuring the safety of the American people. In any case, they are defensible both in principle and in law—and, as I say, they continue to be a matter of lively public debate in which all are invited to join.

For some, however, this is evidently not enough. It is no doubt a measure of how adroitly Muslim Americans have adapted to our general ethos of entitlement that they and their representatives should so uninhibitedly denounce even the most timid expression of concern about the Islamist danger as "Islamophobia," or as an infringement on their "rights." But they do, and this has proved an effective tactic of intimidation. Considering the lengths to which our country has gone to accommodate Muslim requirements, it is also sheer effrontery. It is of a piece with the effrontery of some Muslim Americans who, having apparently decided after September 11 that the time has come to leave our shores, have taken pains to let us know that it is all our fault.

"I don't want to live in a police state," was the thought expressed to a *New York Times* reporter by a native of Jordan who, having resided here for thirty-three years, and having won many civic awards in his Orange County, California, town, was now contemplating a reverse voyage. "Every family is talking" about going home, confirmed the managing editor of an Arabic paper in Anaheim. "I'm scared to death," said a third, a Palestinian, who had been here for eleven years. Although the reporter was careful to record that, for every Muslim who talked of leaving, at least "an equal number" were determined to stay—"After all, they argue, where else are the economic opportunities so promising and . . . the freedoms so plentiful?"—the picture he painted was chilling, if not for the reasons stated by his interviewees.

Sixty years ago, in another war, a whole resident population fell under suspicion of harboring possible collaborators with the enemy. In an action that has been a source of controversy to this day, and for which our government has formally apologized, substantial numbers were placed in internment camps for the duration of the war. I am speaking of course about ethnic Japanese living in specified military areas during World War II. Do we need to remind ourselves that, to say the least, nothing remotely approaching this threatens Muslim Americans today? Or that, as far as discrimination and bias are concerned, today's social attitudes are about as distant from those of the 1930s and 1940s as can be imagined? Do we need to remind ourselves of the many Japanese-American citizens who, far from averring that America meant "nothing" to them, volunteered and served with bravery in the armed forces of the United States during the war itself?

It is not only here at home that the United States has compiled a praiseworthy record of succor and accommodation where Muslims are concerned. Around the world, we have intervened repeatedly during the past decade in behalf of Muslim interests. We defended Kuwait and Saudi Arabia from Saddam Hussein; we stopped wholesale persecutions of Muslims in Bosnia and Kosovo; we brought assistance to the Muslim nation of Somalia; we were long the biggest provider of food in Afghanistan itself, and now we have liberated that country from the boot of the Taliban. Yet we have often been rewarded for our efforts with petulance, double-dealing, resentment, hatred—and terror. That large parts of the Muslim world remain sunk in economic and social degradation is a fact for which we are assigned the blame.

Too often, we have tacitly accepted a share of that blame, tacitly behaved as if we needed to ask forgiveness for the weakness and backwardness, the corruption and evil, that others have brought on themselves and for which they are solely responsible. If the Islamic world is ever to experience the uplift it has demanded, all this will have to change—on both sides. They will have to cease rejecting Western civilization and instead begin to study it; we will have to cease indulging ourselves in guilt and instead, as the writer Shelby Steele has finely put it, "allow the greatness of Western civilization to speak for itself."

What do I *not* mean by this? I do not mean—I would be the last person in the world to advocate—the adoption by others of those grosser aspects of American popular culture to which the unfortunate imam from New York alluded. (Though some of those grosser aspects seem to have exercised an irresistible appeal to at least some of the September 11 hijackers, who were seen drinking in a Florida strip joint on the eve of the attack.) Nor do I expect the wholesale importation even of such indispensable staples of our own lives as individual property rights; freedom of expression, assembly, and religion; and the equality of men and women, though these have going for them a peerless record of empirical success and must also, for many Muslims, constitute a powerful lure.

It is not for me, or for any of us, to tell others how to arrange their domestic affairs. What we can tell them is something else: that the habit of self-criticism, which some in the West have admittedly made into a self-destructive fetish, also happens to be the one irreplaceable engine of human progress. If, as we have been assured, moderates really do outnumber extremists in the Muslim world, if they really do not wish to relegate themselves to the extremists' enclaves, and if they really do aspire to enjoy the benefits of a free society, then they will have to stand up and begin the arduous work of reconstruction from within by criticizing, criticizing, criticizing: wresting the souls of their children from the clutches of self-pity and resentment, taking on the extremists at every point, defining their patrimony anew, and trusting to their convictions, their faith, and the God of history for a vindication that may yet be theirs.

What the rest of us can do to help is, paradoxically, to resist: resist the temptation to excuse the reality of Islamist hatred; resist the impulse to make alliances of convenience with forces of intransigence; resist what Bernard Lewis calls the "anxious propitiation" of dictators and tyrants; resist those who tell us that their fate is our fault, their destiny our responsibility. That, and this, above all: encourage and support our true friends, and finish what we've started.

CHAPTER 4

THE CASE
OF ISRAEL

IT "MAKES A SANE MAN GO MAD," lamented Saud al-Faisal, the foreign minister of Saudi Arabia, speaking in early November 2001 to the *New York Times*.

Quite apart from the particular subject he was speaking about, the foreign minister—like almost everybody else who holds power in that country, a prince of the royal family—had ample cause to fear for his mental equilibrium. How, a reader of the *Times* might wonder, would this personage ever be able to explain himself and his country to the American people? After all, Saudi Arabia had been named as a member in good standing of our coalition against global terror; how could that possibly be reconciled with the fact that the world's chief terrorist, Osama bin Laden, was the moneyed scion of a prominent Saudi family, or with the fact that, as it quickly emerged after September 11, fifteen of the nineteen suicide hijackers were themselves Saudi nationals?

Nor was that the only possible source of contention between Saudi Arabia and the United States. Staggering amounts of money from the corrupt regime the prince served had been going to back radical Islamic militancy around the globe. The (public!) schools in Saudi Arabia were indoctrinating children with anti-Christian, anti-Jewish, and anti-Western hatred. To add insult to injury, even as it took its place in the U.S.-led coalition against terror, the kingdom had refused to join ninety-four other countries in complying with our request to supply American customs agents with advance lists of passengers on U.S.-bound flights. Was this the behavior of a friend?

Bizarre as all this was, however, it was not what the *Times* had chosen to inquire about in its interview with the foreign minister, or what seemed to be disturbing his serenity. No, what interested the paper, and what he confessed tipped his brain toward "madness," was something else altogether: namely, the perceived unwillingness of the American government to compel the democratic state of Israel to sign a "peace agreement" with the radical Palestinian forces bent on its destruction. "The thing that is so sad," concluded the prince lugubriously, "is that what is needed to make peace is very little."

TO BE FAIR TO THE PRINCE, HE was not the only person after the terrorist attacks on September 11 who pretended to be puzzled, or who really was puzzled, by our longstanding support of Israel. Many wondered aloud whether all our troubles might be related to this one circumstance. Americans were pretty sure that we had not done anything directly to bring on the attack, but we were perhaps less certain that we had not somehow contributed to it indirectly through our connection with Israel. For here, some said, was a country whose harsh treatment of the rights of the Palestinian Arabs was a source of inflammation throughout the Arab and Muslim world, and even Osama bin Laden, in his litany of complaints directed mostly at us, had made room for a promise to undo Israeli "aggression" against the Arabs. Did it not follow that pressuring the Israelis to satisfy the just grievances of the Palestinians might bring peace at last to the Middle East and, by removing a prime "root cause" of Muslim rage, help assuage the sentiments that had led to al-Qaeda–type terrorism?

That was the moderate statement of the position. To some who espouse this line of thinking, its appeal rested as well on a sweeping judgment, amply fed by sources in the Muslim world, that Israel had actually stolen Palestinian land and had used "brutal," "racist," or "Nazi-like" means to maintain its control over the hapless Palestinian population in gross violation of international law and morality alike—and had implicated the United States in this infamy. Thus the extreme leftist Noam Chomsky: "Bin Laden . . . is outraged by longstanding U.S. support for Israel's brutal military occupation, . . . the killings, the harsh and destructive siege over many years, the daily humiliation to which Palestinians are subjected."

As it happens, every single point in the above recitation is a canard. But the general proposition has refused to go away, and it even survived the riot of terror against Israeli civilians by Palestinian and Islamist groups in the late fall of

2001. In addition, for a period during the start-up of our own military campaign, its surface plausibility was enhanced by mixed signals coming from our government as we put together the coalition to eradicate terrorism. In doing so, we were seeking to include not only Saudi Arabia but a number of other Arab and Muslim countries that were either officially defined by us as terrorist states themselves or were known to harbor and support terrorist organizations, including al-Qaeda–type groups like Hezbollah, with much American blood on their hands. At the same time, we pointedly excluded Israel, our one democratic ally in the Middle East and the country that has not only been the principal target of Arab and Islamic terrorism but has had the longest experience fighting it—and whose citizens felt with the most profound grief and sympathy our own devastating losses on September 11 even as Palestinians and others in the Arab world were openly celebrating them. (No one who saw the ghastly footage of Arab rejoicing at our misfortune, before it was yanked off the air, will ever forget it.) As if this were not enough in itself to sow moral confusion, we also declared our determination, seemingly as part of our selfsame war on terror, to help create a Palestinian state headed by Yasir Arafat, the one world leader with the distinction of being simultaneously a terrorist himself, an instigator of terrorism, and a supporter and harborer of terrorists.

So, in that limited sense, one did not have to be a cynical politician like Prince Saud al-Faisal to have felt rightly puzzled. Let me see if I can sort out this mess.

DOES HATRED OF ISRAEL (the "Little Satan") fuel hatred of America (the "Great Satan"), such that appeasing the one hatred could dispel the other? Let us begin with Osama bin Laden, who was nothing if not forthright about the reasons behind his hatred of America.

True, bin Laden's list of "grievances," like a comparable list drawn up by Adolf Hitler during his own rise to power, was an ever-expanding one, and we should not have been surprised to see entries on it like Palestinian deaths in the *intifada* that began in the fall of 2000, or the dispute over the territories occupied by Israel after the Six-Day War of 1967. But these were actually late additions, afterthoughts, to bin Laden's basic agenda, which was really aimed at toppling the insufficiently radical Saudi monarchy and other deficient Muslim regimes, gaining access to nuclear weapons, and prosecuting a worldwide war against the "infidel" and "decadent" West. It was not the situation of the Palestinians that first inspired bin Laden's maniacal zeal—or that prompted quaking Saudi princes and others to throw money at their erstwhile protégé in an effort to buy him off and forestall what he intended for them.

The president of Egypt, Hosni Mubarak, was another who argued that most terrorist incidents in the world could be traced to the festering Israeli-Palestinian dispute. Yet it was not the situation of the Palestinians that had driven Islamists associated with Osama bin Laden to attempt to assassinate this same Mubarak in 1995, or that motivated the terrorist massacre of Swiss and Japanese tourists at Luxor in 1997. And if the Arab-Israeli conflict is what fuels terrorism, how did the conclusion of a *peace* deal between Israel and Egypt in 1978 lead directly to the murder of then-president Anwar Sadat in an Islamist plot prominently featuring the man who became Osama bin Laden's closest associate? Was it the plight of the Palestinians that drove a columnist in one government-sponsored Egyptian daily to warn, shortly after September 11, that the "Statue of Liberty in New York harbor must be destroyed. . . . The age of the American collapse has begun"?

In short, whatever the connection between the hatred of Israel and the hatred of America—and I do not for an instant deny the connection—it is not the simple one-way street posited by those on Left and Right alike who have looked for a convenient solution to the challenge before us. In fact, one could more plausibly argue, as Norman Podhoretz did in the *Wall Street Journal*, that the connection ran the other way: that "the hatred of Israel is in large part a surrogate for anti-Americanism," and that "if Israel had never come into existence, or if it were magically to disappear, the U.S. would still stand as an embodiment of everything that most of these Arabs consider evil."

What about the idea that, for reasons either of realpolitik—that is, preserving the antiterror coalition—or of justice, we ought to do what we can anyway to bring peace to the Middle East by resolving the dispute between Israel and the Palestinians? A coalition is a very fine thing, and so is resolving the Arab-Israeli conflict; but they are also separate things, and each presents difficulties.

As I remarked earlier, the usefulness of any coalition depends on the degree to which it serves our present purposes and not the possibly divergent purposes of our partners—and also on the degree to which we are getting from those partners what we need. It was thus unfortunate that we introduced a lack of clarity about our purposes when we welcomed into the coalition known sponsors of terror, and even sat by quietly and allowed one of them, Syria, to take a seat on the Security Council of the United Nations. We may also have compromised whatever sympathy we could have counted on among ordinary, peace-loving Muslims when we made common cause with the dictators and tyrants they fear and already suspect us of coddling. I'm not arguing that coalitions need be made up only of friends (though I am against *excluding* friends, especially valuable ones). I am arguing for discrimination, for toughness rather than propitiation, and for close attention to moral and political messages.

As for resolving the dispute between Israel and the Palestinians, that is a noble goal—none nobler. But President Bush came into office with a much nobler and more accurate understanding: that, being fundamentally asymmetrical, this dispute is insusceptible of solution by means of outside intervention, except at the cost of irreparable harm both to Israel and to the long-term interests of the United States—and, I would add, to the Palestinian Arabs themselves, who would thereby be consigned to go on suffering under the despotic heel of Yasir Arafat's Palestinian Authority. For a little while, in the early stages of the war, it seemed we might be losing sight of this true understanding, bending to pressure by our coalition partners to distance ourselves from Israel and, at a moment of high tension and for the first time in our history, to call openly for the establishment of a Palestinian state. That would have been to betray both realpolitik and principle alike, forfeiting our position of leadership in the coalition we had assembled and repudiating a friend into the bargain.

How is the dispute between Israel and the Palestinians asymmetrical? It is asymmetrical because one side in the conflict wishes to live in peace and has never wished otherwise, while the other side is bent not on living in a state next to Israel but in an Arab state replacing Israel.

THAT THE ISRAELI PEOPLE ARE bone-weary of the war that has been waged against them since before the day of their nation's birth in 1948, and that they are almost recklessly desirous of peace, is a proposition that to me is so self-evident as to need no demonstration. To even the most skeptical observer, this proposition should have been confirmed when, at the very height of the violence directed against their civilian population in the fall of 2001, poll after poll registered majorities of Israelis in favor of continued negotiation with Yasir Arafat, up to and including an agreement on a Palestinian state.

Arafat's own, contrasting ambitions should also have been clear. A year earlier, in talks convened by President Clinton, Israel's prime minister, Ehud Barak, had handed the terrorist leader virtually everything he had ever demanded. Arafat turned it down, forgoing the opportunity to build his state in favor of launching an *intifada* that made the streets of

Israel run red with the gore of innocents—the equivalent in Israeli dead of twenty thousand Americans over the course of a single year.

Nor was there any mystery in Arafat's move, at least to eyes unclouded by what the writer Mark Helprin has rightly dubbed the "misconception, delusion, conformity, and cowardice" that made up the nearly decade-long Oslo peace process. From the beginning, the Palestinian strategy was, and still is, to provoke outrages to which Israel has had no choice but to respond with stringent countermeasures, thereby supplying endless footage of Palestinian casualties and eliciting fresh pressure by the West to accommodate ever-escalating Palestinian demands. Israel's commitment to peace stimulated in the Palestinians not a spirit of reciprocal amicableness but an appetite for more—for everything—defined publicly as not only a Palestinian state in the "disputed territories," with Jerusalem as its capital, but an unlimited right of return for all Arab refugees to Israel proper, which would effectively drown the Jewish state in a demographic deluge.

Taken all in all, given Arafat's long record of prevarication, duplicity, and incitement, it would be very difficult to argue that his ultimate goal is anything other than the destruction and replacement of Israel by any and all means possible. The official emblem and stationery of the Palestine Liberation Organization (PLO), headed by Arafat, feature a map of the entirety of what is to be "liberated": not just the "disputed territories," but all of present-day Israel. The Fatah wing of the PLO, its so-called moderate faction, has as its insignia a shield bearing, again, the entirety of what is now

Israel, surmounted by two fists holding rifles with a hand grenade beneath. In textbooks for Palestinian children, maps do not show Israel at all, only an area defined as the former and future Palestinian homeland. As Arafat is fond of saying in his speeches to Arabic-language audiences, "we will wave the flag of Palestine, Allah willing, over the walls of Jerusalem . . . whether someone likes it or not, and whoever does not like it can drink the water of the Dead Sea."*

Privately, then, and when they are not speaking for Western ears, it seems to me there is little or nothing to distinguish the program of the "moderate" PLO from that of certified terrorist groups like Hezbollah or Hamas or Islamic Jihad. The differences involve tactics, the end goal remains the same—and even tactically, both Arafat's own Fatah wing of the PLO and his armed militia, the Tanzim, have been implicated in overt terrorist acts, fully justifying Prime Minister Ariel Sharon in classifying the Palestinian Authority as a terrorist organization in the same category as al-Qaeda.

In one way or another, whether militarily, politically, or demographically, whether by negotiation or confrontation or terror, the end of all this activity looks to be the final dissolution of the Jewish state and the establishment in its place of an Arab Palestine "from the [Jordan] River to the [Mediterranean] Sea." In exactly those words was it defined by one famously moderate spokesman of the Palestinians, Faisal al-Husseini, in

---

*Here and elsewhere in this chapter, as well as in the previous chapter, I have made use of the invaluable translations from Arabic and Muslim media made available by the Middle East Media Research Institute (MEMRI).

an Arabic-language interview before his death last year. And lest anyone think Israel is so large and mighty that it can easily triumph over any such intention, let us remember that it is in fact very small—smaller than the state of Maine or Lake Michigan, smaller by far than Egypt or Iraq or Syria or Jordan. A small nation, wrote the Czech novelist Milan Kundera, is "one whose very existence may be put into question at any moment; a small nation can disappear, and knows it."

I DO NOT MEAN TO RELATE at length the story of the Zionist movement and the founding of the Jewish state, though to anyone who cherishes freedom and the drama of human redemption that story is thrilling beyond measure—in some ways, as the historian Paul Johnson has written, the quintessential story of our age. ("In the last half-century," Johnson observed in 1998 on the occasion of Israel's fiftieth anniversary, "over 100 completely new independent states have come into existence. Israel is the only one whose creation can fairly be called a miracle.") Nor need we linger over what Johnson rightly terms the "supreme folly" of Arab rejectionism, exercised in full ferocity in the multinational invasion of the infant Israeli state in 1948 and, ever since then, constituting the primary engine of the agony known as the Arab-Israeli conflict. Once again, the bottom line is clear: To the Jews' dream of peaceful integration the Arab nations have repeatedly counterposed their own dream of Jewish extinction, one that only the force of Israel's arms and its determination to survive have turned aside till now.

Still, it might be worth reminding ourselves briefly of how the so-called disputed territories—the West Bank and Gaza—came into existence. They were certainly not "stolen." Before being acquired by Israel in a defensive war in 1967, they had been occupied since 1948 by, respectively, Jordan and Egypt. Jordan's title to the West Bank (originally designated by the British as part of the Jewish National Home) had never been recognized in international law, and in any case neither Jordan nor Egypt nor any other party had ever claimed these territories as a Palestinian homeland. Neither had the Palestine Liberation Organization (PLO), founded in 1964, which was no surprise since back then, too, the only homeland the PLO was intending to "liberate" was the land of the state of Israel. As for the 1967 war itself, it had nothing to do with the situation of the Palestinian Arabs; as Gamal Abdel Nasser, the president of Egypt, put it with complete frankness, the "aggression" the Arabs sought to undo in provoking that war was the existence of Israel itself.

To be sure, the Arab states had made much of the hapless condition of the Palestinian Arab refugees who had fled their homes in the 1948 war that accompanied the birth of Israel. But these states had also ensured that the refugees' condition would *remain* hapless by refusing to absorb and resettle them as the Israelis had done with an equivalent number of Jewish refugees who had fled or been expelled from their homes in the Arab world during the same war. Almost twenty years later, in the immediate aftermath of the Six-Day War of 1967, Israel was eager to negotiate a return of the land it had conquered in exchange for peace and diplomatic recognition by the Arabs, and there was nothing to prevent those territories from becoming an absorption point for refugees or indeed the locus of a Palestinian state. The answer, delivered at a meeting of all the Arab states in Khartoum in August 1967, was swift and plain. It was the famous "Three No's": no peace, no recognition, no negotiations.

Thus, in the face of Arab adamancy, began the Israeli occupation and administration of the West Bank and Gaza, and thus, too, began the myth of Israeli "intransigence." Translated into English, the only thing this could mean was the continued determination of the Israelis to resist Arab plans for their elimination. The real intransigents were the Arabs.

Has anything changed in the thirty-something years since then? Yes, and no. The PLO's terror campaign, the model of every other such campaign in contemporary history, turned out to be wildly successful, and in a matter of years—years that saw repeated airplane hijackings, the murder of Israeli Olympic athletes, terror bombings, and blackmail on a global scale—Yasir Arafat emerged with the respect and recognition of much of the world. No less successful was the propaganda effort, undertaken with the aid of the Soviet Union and the complicity of the international Left, to paint the Palestinians as the victims of Jewish aggression and the Israelis as a species of new Nazis. This effort reached its high point, or its nadir, in 1975, with the U.N. resolution equating Zionism with racism. By the early 1990s, with the ardent collaboration of our own government, the Israelis were negotiating at Oslo with their sworn enemy, and essentially suing for peace. That they have not yet achieved it is owing only to their stubborn hope of getting something, anything, in return.

IN THE MEANTIME, AND IN THE interstices of this never-ending hostility, let us not forget who the Israeli people are and what they have accomplished. They have forged a thriving, successful, brilliantly variegated, and racially polychrome society, one that another admiring historian, Conor Cruise O'Brien, has called "so democratic as to be almost unworkable." That vibrant society is a standing reproach to the drab tyrannies that have immiserated and enchained the populations of the neighboring Arab states, and from which the luckless inhabitants of those states have so far failed to extricate themselves. In the entirety of the Middle East, there is but one democracy, one country where all—Jews, Christians, Muslims—can be citizens, all can vote, all can be and are represented in a democratically elected parliament.

What about the Arab inhabitants of the disputed territories under Israel's administration? For all the undeniable tribulations of occupation, it is a matter of record that they, too, have enjoyed a higher standard of living and a broader range of rights than their brethren anywhere in the region can dream of. I am hardly the first observer to note—many Palestinians have themselves testified—that their intimate experience of a working democracy is what has exacerbated their feelings of discontent far more sharply than have the alleged "brutality" and "humiliation" of Israeli control. In the summer of 2000, when it began to appear likely that a deal was near between Israel and the Palestinian Authority, Israeli government offices were flooded with Arab Jerusalemites seeking citizenship before "their" state caught them once and for all in its clutches.

It is around this point that we can begin to grasp more fully the real connection between Israel and us, and between hatred of Israel and hatred of us. What most Palestinians want, I have to believe, is what they say they want: justice and freedom and peace. But there is no place in the Arab world where those blessed entities are to be had, and precious little prospect of their coming into being in a Palestinian state. Similarly, I have to believe that most Palestinians value life, and want to live, but their own culture glorifies war, and armed struggle, and death, including the death of little children on suicide missions.

Left to fend for themselves, in their own world, among their own Arab brethren, the Palestinians would be just as they always were in history: out of luck. To what, then, are their spokesmen appealing when they demand sympathy for the Palestinian cause? Where are justice and freedom and peace not just empty words but realities? Where is human life the supreme value? The answer is obvious, and so is the hypocrisy, and the pathos, behind the demand being made of us. As the writer Norah Vincent has coldly but truly put it, "If it weren't for our (and Israel's) cultural commitment to tolerance and the rule of law, to the use of violence only in self-defense and to the reaching of diplomatic solutions, the Palestinian people would have no cause at all. They would not exist."

And what if their cause should triumph? "Do you imagine," Vincent asks, that the new state of Palestine "would be
anything other than a repressive dictatorship bent on crushing its God-given enemies?" And "do you really suppose
there would be any Jews left to protest?"

That is why I continue to say that the burden is not on
Israel to make peace; the burden is on the Palestinians to
renounce their terrorism and on both the Palestinians and
the Arab states to renounce their murderous intentions
toward Israel. The day they make their own inner peace with
the existence of a sovereign Jewish state in their midst will be
the day the Arab-Israeli conflict ends.

And where do we come in? This, too, is a long story—
one that does not in the least conform to the conventional
wisdom that our conduct in the Middle East is one-sided or
that our friendship with Israel is eternal. Although we recognized the Jewish state on its establishment, our policy has
been shaped from the start by our role as the successor to
the British and French in maintaining order in that part of
the world, by our interest in containing Soviet power in the
long period of the cold war, and by our economic depend-
ence on Arab oil. Not until the 1967 war, really not until the
1973 Yom Kippur war, would it have been truthful to
describe us as friendly to Israel—before then we were a supplier of arms to the Arabs, not to the Israelis—and even in
the last decades we have consistently pressured Jerusalem at
every turn to "solve" its conflict with the Palestinian Arabs
by any means possible. We have tried, in short, to be evenhanded, not infrequently at the cost of recognizing that to

be evenhanded in an inherently asymmetrical contest is to favor Arab intransigence and irredentism.

Here is the warning of a professor of politics in the *New York Review of Books*: "[O]ur efforts to eradicate terrorism will go for nothing" unless we cease providing "cover" for Israel's unconscionable treatment of the Palestinians. This professor has things backward: Our efforts to eradicate terrorism will go for nothing if we abandon Israel in its struggle to exist by failing to appreciate that others have not ceased trying to annihilate it.

Besides, the "cover" that we, the American people, provide for Israel has very little to do with politics, or with policy in the narrow sense. It has to do with shared human values—the values in whose name we have undertaken our "efforts to eradicate terrorism" in the first place. Over and over again, ordinary Americans have indicated their appreciation of this elementary fact, as well as their instinctive comprehension that, when it comes to the Middle East, the issue at stake is nothing less than Israel's very existence.

In mid-October 2001, after we had formed our coalition and duly excluded Israel from it so as not to offend our Muslim partners, a stunning 72.9 percent of us in a national McLaughlin survey strongly favored our continued support of Israel. No less stunningly, fewer than 10 percent thought pressure on Israel would put an end to anti–U.S. terrorism, while almost 62 percent held the opposite view—namely, that pressuring Israel to surrender territory or divide its capital city of Jerusalem would reward terrorism and incite further attacks.

Only if we were to abandon these solid understandings, and the Israelis with them, would our efforts truly "go for nothing."

I MUST INTRODUCE NOW A SUBJECT that I have so far touched on very fleetingly, but that likewise bears crucially on the connection between Israel and us. If it is true that "they" hate Israel because they hate us, and if it is no less true that they hate both Israel and us for the same reasons, it is also true that they hate Israel especially and most particularly on other grounds. For the sake of our own sanity and moral health, we need to confront this hatred, which demands an unblinking and uncompromising response.

It would be almost impossible to overstate the intensity of the particular hatred of which I speak, whose plain name is anti-Semitism. To many Westerners, the recrudescence in the post-Holocaust period of this oldest disease of our own civilization is indeed almost too horrible to contemplate, which may be why we have tended to turn our faces away from it. But the fact is that everywhere in the Muslim world, most saliently in the Middle East but also elsewhere, far afield of the Arab-Israeli context, it is a fierce and burning fire.

Burning, I hasten to add, long before the onset of the latest *intifada*, long before the Israeli occupation of the West Bank and Gaza, long before any of the "provocations" routinely cited to justify Arab grievances, including the establishment of the state of Israel itself. In 1929, Arab terrorists murdered Jews all over Palestine; in that year, a political pamphlet proclaimed: "O Arab! Remember that the Jew is your strongest enemy and the enemy of your ancestors since olden times." Throughout World War II, the supreme religious leader of Palestine's Muslim Arabs, the grand mufti Haj Amin al-Husseini, eagerly lent his energies to Hitler's cause and especially to Hitler's genocidal program against the Jews. The mufti's example would be eerily echoed almost a half-century later, in the Gulf War of 1991, when Yasir Arafat made common cause with Iraqi dictator Saddam Hussein—the same Saddam Hussein who only a year earlier had promised to "let our fire burn half of Israel."

Murderous anti-Semitism in the Arab world is not occasional or peripheral, it is constant and central. It is sounded at every level, from the pronouncements of high government officials to the rulings of Islamic religious authorities, from the writings of lawyers and journalists and university professors to the effusions of radio and television stars, and in every medium of culture high and low. In the Arab world, wrote Bernard Lewis in 1986, "The demonization of Jews goes further than it had ever done in Western literature, with the exception of Germany during the period of Nazi rule."

Arab anti-Semitism did not fade with the onset of the Oslo peace process in 1993; on the contrary, it increased—tenfold. This in itself confounds the theory that what fuels Arab hatred is Israeli belligerence; what fuels Arab hatred is Arab hatred. Nor was the Oslo pattern a new one. In 1978, Israel concluded a formal peace with Egypt. Today, Egypt may be the single most prolific disseminator of anti-Semitic hate literature in the world. As the Italian journalist Fiamma

Nirenstein has documented, not only is the notorious nine-teenth-century forgery *The Protocols of the Elders of Zion* published and distributed widely in Egypt, but government-sponsored newspapers routinely purvey the medieval libel that Jews use the blood of non-Jews in baking Passover matzo; Egyptian news sources have "repeatedly warned that Israel has distributed drug-laced chewing gum and candy, intended (it is said) to kill children and make women sexually corrupt"; and a leading columnist in a government-sponsored newspaper can write openly in praise of "'Hitler, of blessed memory, who . . . took revenge in advance on the most vile criminals on the face of the earth.'" Last year, government-run radio stations in Egypt made a number-one hit of a song entitled "I Hate Israel."

Not to be outdone in vituperation are the Syrians, whose former defense minister in 1983 wrote a book entitled *The Matzo of Zion*—soon, we are told, to be a major motion picture—and whose current president, Bashar al-Assad, welcomed Pope John Paul II to his country last summer by delivering, at the airport ceremony, a blistering tirade on the perfidiousness of world Jewry. A textbook for Syrian tenth-graders instructs: "The logic of justice obligates the application of the single verdict on the Jews from which there is no escape: namely, that their criminal intentions be turned against them and that they be exterminated."

A prime locus of anti-Semitic propaganda has been, of course, Yasir Arafat's Palestinian Authority. In blatant violation of commitments undertaken at Oslo, Palestinian schools and camps, radio and television, newspapers, and every organ of official opinion have spread implacable and virulent hatred of Jews and exhortations to murder and annihilate them. One need look no further than this indoctrination in hatred, Fiamma Nirenstein writes, to understand how children grow up wanting to be suicide bombers, or to understand how, even after September 11, over three-quarters of Palestinians continued to approve of this ghastly form of sanctified murder and child sacrifice. "There is no shortage of willing recruits for martyrdom," reported Nasra Hassan in *The New Yorker* in late November, quoting a leader of Hamas: "'Our biggest problem is the hordes of young men who beat on our doors, clamoring to be sent.'"

"Blessings on whoever has put a belt of explosives on his body or on his sons and plunged into the midst of the Jews," declared Sheikh Ibrahim Madhi in a sermon on official Palestinian television in June 2001. Quoting this sort of thing, Andrew Sullivan has identified "pathological anti-Semitism" as a central component not only of the Arab-Israeli conflict but of the whole "irrational, lethal movement stirring many people across the globe in a call to mass murder." So twisted and all-encompassing is this pathology that some Muslims have held the Jews themselves responsible for the bombing of the World Trade Center. As noted previously, the former imam of the Islamic Cultural Center of New York was among them; so, too, were a plurality of Egyptians and, in one *Newsweek* poll, 48 percent of Pakistanis.

These are the people, and these are the leaders, with whom Israel is being told to sit down and engage in a peace process—such a "very little thing," after all.

THE FANATICAL ANTI-SEMITISM of large segments of the Arab world, notes Andrew Sullivan, is "not a revelation," and yet he himself, speaking for many of us, had long tended to ignore or discount it. "Why," he asks, "did I look the other way?" He cites the usual grounds: that "these are alien cultures and we cannot fully understand them, or because these pathologies are allied with more legitimate (if to my mind unpersuasive) critiques of Israeli policy." His conclusion is very sharp:

> We in the West simply do not want to believe that this kind of hatred still exists; and when it emerges, we feel uncomfortable. We do everything we can to change the subject. Why the denial, I ask myself? What is it about this sickness that we do not understand by now? And what possible excuse do we have not to expose and confront it with all the might we have?

These are piercing questions. For me, they rang with all the greater urgency when, late last fall, I began to read about anti-Semitic statements and innuendo that were emanating not from the Arab or Muslim world but from, of all places, Western, democratic Europe. Looming before my eyes was not a "denial" of this hatred but an open embrace of it in the most advanced circles of society.

In France, the respected left-wing weekly *Le Nouvel Observateur* reprinted a wholly unconfirmed report to the effect that Israeli soldiers routinely raped Palestinian women in order to induce their families to murder them for having stained the family honor. In Belgium, the media were being even more vicious than the French in their attacks on the government of Israel, while the judiciary was busy preparing for a possible trial of Ariel Sharon in connection with an event in Lebanon twenty years earlier that had been investigated exhaustively by an independent Israeli commission. In Germany, the publisher of the influential weekly *Der Spiegel* suggested that Sharon's attitude toward the Palestinians bore comparison with Hitler's toward the Jews.

And in England? There the idea took hold that Israel, or "the Jews," had succeeded in embroiling the West in an unnecessary and unwanted conflict with the Islamic world, and that something had to be done about it. The well-known novelist A. N. Wilson wrote a column asserting that, thanks to ceaseless warmongering by the state of Israel, what must now be put into question was its right to exist at all. (Is there any other polity in the entire world, no matter how barbarous, about which a Western intellectual would permit himself to say such a thing?) According to an article by Barbara Amiel in the weekly *Spectator*, since September 11 the open expression of anti-Semitism had "become respectable at London dinner tables"; Amiel reported the words of a liberal member of the House of Lords: "Well, the Jews have been asking for it and now, thank God, we can say what we think at last."

One can speculate endlessly about what lay behind these outbursts and others that were still to come. In drawing up a possible list of motives, another British writer, Melanie Phillips, named such factors as the poisoned culture of victimhood, upper-class self-loathing, and the Left's "abiding hatred of Israel, America, and . . . Western values." Noting that the European Union, of which Britain is a member, was "the largest supplier of funds" to Yasir Arafat's Palestinian Authority, Phillips speculated pointedly that the British, our main partner in the fight against global terror, had now "succumbed to the very prejudices" that lay behind that terror.

This conclusion may be too sweeping, and also, one hopes, will be proved wrong. At least one leading British politician, Iain Duncan Smith, the new head of the Conservative party, came out publicly to accuse the BBC and others in the British media of anti-Israel bias and of conferring legitimacy on terrorist organizations like Hamas and Islamic Jihad.* But there is no gainsaying the awful power of anti-Semitic prejudice that made itself felt in Europe in the wake of September 11, or rather in the wake of our vigorous response to September 11. Nor is there any gainsaying the historically potent role played by this prejudice in motivating no less powerful impulses of both scapegoating and appeasement.

THAT IS WHY OUR BEHAVIOR with regard to Israel is so crucial. And it is also a reason to have a quick look back at an insufficiently noticed event that, concluding less than a week before September 11, formed a macabre prologue to it.

That event, held in Durban, South Africa, was the United Nations–sponsored World Conference Against Racism, Racial Discrimination, Xenophobia, and Related Intolerance.

---

*Our own media, print and electronic alike, are hardly blameless in this respect. The Committee for Accuracy in Middle East Reporting in America (CAMERA) keeps a meticulous record.

It would have been more accurately titled a world conference *of* racism, discrimination, xenophobia, and related intolerance, for all those evils were on abundant display at Durban, where they were directed mainly at two objects. Durban was, in fact, an orchestrated carnival of anti-Israel and anti-American hatred.

Especially prominent was the singular version of racism and xenophobia known as anti-Semitism. Bad enough were the official proceedings, which featured speeches by assorted dictators attacking Israel as a racist state. More grotesque still was the parallel conference for representatives of nongovernmental organizations. There, as the analyst Arch Puddington subsequently reported,

> Delegates from Islamic countries, assisted by a rabble bused or flown into Durban for the occasion, chanted anti-Israel slogans, handed out the vilest sort of anti-Jewish literature, and sponsored anti-Semitic marches and rallies at which participants were adorned with T-shirts bearing anti-Zionist messages. In one incident, an Egyptian lawyers' group distributed a pamphlet featuring cartoons that might have come straight out of [the Nazi newspaper] *Der Stuermer*, with hooknosed Jews grinning over the blood-soaked bodies of Palestinians. Another leaflet had Hitler saying, "If I had not lost, Israel would not exist today."

When they were not assaulting Israel and Jews everywhere, the tribunes of some of the world's most oppressive and murderous regimes were turning their talents for demonization on *us*, in language that differed only in its particulars from the wholesale indictment of Israel.

It was therefore, in my judgment, one of America's proudest moments, one of America's most *significant* moments, when the Bush administration responded to this twinned manifestation of evil by announcing that we were walking out. It was a moment fully in tune with our magnificent action twenty-six years earlier in standing against the U.N. infamy that was the Zionism-racism resolution.

To be sure, the vilification at Durban increased when we withdrew, amplified now by the sanctimonious regrets of some of our European allies. But the question is not whether we were justified in doing what we did; our withdrawal, as Puddington writes, "was one of the few principled acts in the whole grotesque affair." The question is why other democracies did not join us, but instead contented themselves with negotiating slightly more temperate language from their tormentors before pronouncing themselves well satisfied with Durban's results.

A possible and quite terrible answer to that question lay before us months later, when the culture of appeasement had its say in the columns of European newspapers and in the drawing rooms of European society.

THAT CULTURE OF APPEASEMENT itself requires an answer, and it can be of only one kind. Exculpating Arab terror because it is allegedly a response to Israeli occupation, forcing Israel to make concessions to Arab terrorists—these are the bitter fruits of moral obfuscation, of ethnic and religious prejudice, and of blindness. The hatred that confronts both Israel and us has shown that it will not be assuaged so readily. Nor are our enemies upset by our sundry imperfections and failings; no less than any other polity, Israel assuredly has its share of both. What they intend, rather, is for our societies to fall. We must be clear about this, and we must be clear that it is they, all of them, who must fall instead.

"From this day forward," President Bush declared to Congress and the American people in September, "any nation that continues to support terrorism will be regarded by the United States as a hostile regime." To me, at least, that statement is impeccable, and should never be compromised. Our only course is to go on leading as we have been leading, to insist as loudly as we are able that the world focus on the true font of evil, to win our war against that evil, and then to watch the appeasers and the anti-Semites fade ignominiously away.

Israel itself is a model we need to contemplate. It is "a nation that has been fighting terror since the day it was born," observed its former prime minister Benjamin Netanyahu in late October. We may never know how much time Israel bought for us in our decades of negligence, how many American lives it saved by its long-kept refusal to negotiate with or capitulate to terrorist murder and extortion, its resolve to use every means to track down, confront, and undo those who captured and killed its citizens, its crystalline message of defiance. What we do know is that all over the world, especially in the Soviet gulag and in the prisons of

Eastern Europe, captive men gulped great draughts of hope whenever word filtered through of an act of Israeli rescue and punishment: palpable and too rare signals in those dark decades that evil was not everywhere triumphant, everywhere accommodated, everywhere appeased.

I want to put it positively. Our essential human kinship with Israel is something like our kinship with Great Britain, but it is also more particular and less blood-related than that. It is a deep-rooted feeling of linked destinies, a feeling that echoes back to our founding and to the earliest conceptions of the American experiment itself, that new birth of freedom which our fathers identified with the biblical Israelites' emergence from the darkness of bondage. And I believe it also has to do with an understanding, almost religious in nature, that to our two nations above all others has been entrusted the fate of liberty in the world. That—the survival of liberty—is precisely what our efforts to eradicate terrorism are all about.

On July 4, 1976, as we were celebrating the two-hundredth anniversary of our founding, 105 Jewish and Israeli passengers huddled in the darkened terminal of Entebbe airport in Uganda. Their Air France jetliner had been hijacked by terrorists and brought to this place, where the crew and the non-Jewish passengers had been released. Their own lives were now being offered in exchange for imprisoned terrorists. As the final deadline approached, a team of Israeli commandos led by Jonathan (Yoni) Netanyahu appeared out of the darkness, infiltrated the airport, slew the captors, and plucked the terrified victims from certain death. The sole Israeli casualty was the commander himself.

The lightning operation at Entebbe electrified the world. A few years later, this is what the great African-American civil rights leader Bayard Rustin had to say about it:

> I am certain that for years and years to come, perhaps even a thousand years from now, when people are confused and frightened, and they are dispossessed of their humanity and feel there is no way to go except to face death and destruction, someone will remember the story of Yoni at Entebbe. That story will be told to those despairing people, and someone will move into a corner and begin to whisper, and that will be the beginning of their liberation.

There, in a kernel, lies the meaning of Israel for us. I myself am one of tens of millions of Americans who have seen in the founding and flourishing of the Jewish state the hand of the same beneficent God who attended our own founding and has guided our fortunes until now. Keeping faith with the people of Israel in their still unfinished confrontation with evil is, to me, a species of keeping faith with ourselves; breaking faith, a species of self-negation. It is exactly that simple, and exactly that difficult, and exactly that consequential.

CHAPTER 5

LOVE OF COUNTRY

WHAT IS PATRIOTISM? The novelist Barbara Kingsolver, whose nine books include the wildly popular *The Poisonwood Bible*, knew. "Our language of patriotism," she wrote in late September 2001, "is inseparable from a battle cry." In the mouths of those who wield the word, and who "get the most airplay," patriotism is that which "threatens free speech with death. It is infuriated by thoughtful hesitation, constructive criticism of our leaders and pleas for peace. It despises people of foreign birth." And what is the American flag, so contaminated by "the men now waving it in the name of jingoism and censorship"? It is a symbol of war, and of warmongers. "When I look at the flag, I see it illuminated by the rockets' red glare."

Who are the American people, and what is the American nation? A professor at the University of British Columbia knew. The American people, she proclaimed at a feminist conference on October 1, "the American nation that Bush is invoking, is a people which is bloodthirsty, vengeful, and calling for blood."

What, again, does the American flag stand for? The poet Katha Pollitt knew, though in the wake of September 11 she was dismayed to discover that her own daughter, a high school student, had not absorbed the lesson. "My daughter," she wrote in the *Nation* magazine, "who goes to Stuyvesant High School only blocks from the World Trade Center, thinks we should fly an American flag out our window. Definitely not, I say: The flag stands for jingoism and vengeance and war."

AND WHAT AWFUL DISPLAY OF jingoism were these people reacting to? They were reacting, as it happens, to a reaction: to the spontaneous upwelling of national feeling that followed upon September 11, the day of trauma. Quite suddenly, as if in the twinkling of an eye, everything petty, self-absorbed, rancorous, decadent, and hostile in our national life seemed to have been wiped away. Suddenly, our country's flag was everywhere, and stayed everywhere. Suddenly, we had heroes again—and what heroes: policemen and firefighters, rescue workers, soldiers, and civilian passengers who leapt from their seats to do battle with evil personified.

It was true; for weeks and even months after September 11,

partisan political issues seemed to fade in urgency, racial divisions to be set at naught. Cynicism and irony were declared out, simple love of country in. Rock stars who only yesterday had been me, me, me-ing us to distraction fell over themselves to donate their time and their talent and their profits to aid the victims of the attacks, and sang their lungs out for America the beautiful. Even the universities, or at least many of the students at them, rallied around the flag.

And what a wonderful, heart-swelling surprise *that* was, especially to those of us, veterans of the "culture wars" of the last three or four decades, who had kept an alarmed watch over the hardening of divisions among us and the downward course of our country's social and cultural indicators (as I termed them in a number of reports and books). There were moments during those years when even the basic, taken-for-granted unity of the United States, in anything more than a rhetorical sense, was beginning to seem in doubt. But the events of September 11, and the amazing response to them, had stilled all such doubts.

Something in those events, wrote an uplifted Peggy Noonan, "something in the fact that all the different colors and faiths and races were helping each other, were in it together, were mutually dependent and mutually support-ive, made you realize: we sealed it that day. We sealed the pact, sealed the promise we made long ago. . . . We are Americans."

That, however, is precisely what quickly gave rise to a countermovement of America-bashing. Both abroad and at home, some were quick to find *us* to blame for the events of

September 11. Some were quite explicit about it. "We regret the thousands of people who died," said the militant black activist Lorenzo Komboa Ervin, "but we know it is Washington's fault this happened." The world, agreed the novelist Arundhati Roy, has been "laid to waste by America's foreign policy" and its "marauding multinationals who are taking over the air we breathe, the ground we stand on, the water we drink, the thoughts we think"; Osama bin Laden was nothing but the "spare rib sculpted from" that world.

Others were somewhat, but only somewhat, more indirect. The response of a professor of history at New York University to the murders of September 11 was to call for "civil war, class war, whatever, to put an end" not to those who would incinerate innocents by means of hijacked civilian aircraft but "to U.S. policies that endanger us all." The director of the Children and Family Justice Center reminded us that terrorism takes many forms, including "U.S. interventions" abroad. And the House of Bishops of the Episcopal Church pointed elliptically to "the death of 6,000 children in the course of a morning," a death caused not by terrorists but by the "crushing poverty" to which our own affluence stands "in stark contrast"—thus equating deliberate murder with an economic condition to which we may (or may not) have indirectly contributed.

The impulse to blame America was not restricted to the Left. Some evangelical Christians, most notoriously Jerry Falwell but also the columnist Joel Belz in *World* magazine (the nation's fourth-largest newsweekly), the social philosopher Marvin Olasky, and one or two others likewise located the "cause" of September 11 in certain specific sins or sinners among us. In their case, they meant sin literally.

Falwell pinpointed "the pagans and the abortionists and the feminists and the gays," as well as all those "who have tried to secularize America." Belz invoked our "false deities" of "nominalism, materialism, secularism, and pluralism," all of which he saw symbolized in the twin towers of the World Trade Center. More plainspokenly, the singer Charlie Daniels wrote on his website that "We've shaken our fists in God's face for far too long. We have ignored His laws, belittled His son, taken His name in vain until it's almost a national slang word. . . . This road leads to hell." Meanwhile, still on the Right but in a wholly different part of the forest, David Duke, the white racist, contended that what led to September 11 were earlier U.S. bombings of Iraq and Afghanistan—and, of course, American support of Israel.

The ideas of Falwell and the other blame-America types on the Christian Right were roundly condemned by most conservatives, including me—and I have not exactly been reticent when it comes to criticizing the flaws of American society. Falwell quickly backed away, saying quite correctly that "no human being has the knowledge that any act is an act of God's judgment." No less vigorously repudiated was the view articulated by David Duke, although the truth is that few conservatives bothered to take notice of it, and for good reason: Sentiments like Duke's are an anomaly on the Right these days. On the Left, by contrast, they are at home, and are hardly criticized at all.

Another difference, even more significant, is that the few

lonely blame-America figures on the Right, whatever troops they may command among religious conservatives, wield little or no influence over mainstream American culture. On the Left, again by contrast, such people tend to be figures with mainstream respectability—professors, editorial writers, columnists, intellectuals, artists, performers, religious spokesmen. Their influence is large, and it is magnified both by their ability to broadcast their ideas widely and by the receptivity of many educated Americans to some version or other of those same ideas.

Few people would actually subscribe to the extreme formulations of the literary scholar Fredric Jameson, who found "the seeds" of September 11 in the "wholesale massacres . . . systematically encouraged and directed" by the United States against leftists and left-wing causes around the globe. Still, many would sympathize with the underlying notion of American guilt. Few might actually say, with the historian Michael Rogin, that Americans regularly inflict violence on others "on a scale that dwarfs the World Trade Center obliteration and the Pentagon fire," but many say or think such things at, as it were, a discount. And that is problematic enough.

SPOKEN OR UNSPOKEN, WHERE do such attitudes come from? As I said a moment ago, the last few decades have witnessed bitter debates—"culture wars"—that have roiled the surface of American unity, and some of them have shaken us to the core: debates over race and ethnicity, over the role of America in the world, over the viability of our economic system, over the goodness of our society and customs, even over the fundamental legitimacy of our constitutional arrangements. The strife was perhaps at its fiercest during the Vietnam period, when the "responsibility of intellectuals," to use a phrase made famous by Noam Chomsky, a left-wing guru of that age, became coterminous with hostility to America's political leadership and American purposes more broadly. In certain quarters, America itself was spelled with a "k"—"Amerika"—to suggest an identification with Nazi Germany, and in those years there was even open talk of revolution, of tearing down the "system" and starting the entire enterprise anew.

But the sixties themselves, and everything they signify, did not emerge out of nothing and nowhere. The radicalism of the New Left had obvious roots in the 1930s, the heyday of communist and left-wing activism in the United States. And feeding this radicalism was still another phenomenon, one whose historical roots went much deeper, and that is worth briefly recalling, if only to suggest the staying power of what those of us who wish to solidify the sentiments of September 11 are up against.

In the period right after the Civil War, the historian Shelby Foote reminds us, Americans ceased to speak of their country in the plural ("the United States are . . .") and began to speak of it in the singular ("the United States is . . ."). The reason was plain: Like no other event in our history, the Civil War had brought home to every American the cost of irreconcilable division; from then on, we would speak of ourselves, and think of ourselves, as one. Curiously enough, however, it was in those same years that homegrown *anti*-American sentiments also began to manifest themselves with force and articulateness.

In his recent memoir *My Love Affair with America*, Norman Podhoretz recalls being introduced to these sentiments as a young student of literature. In their earliest incarnation, he reminds us, they came in both a left- and a right-wing version; taking root in what came to called the Gilded Age, they then "spread inexorably through the literary culture over the next hundred years." Inspired by muckraking journalists like Ida Tarbell and Lincoln Steffens and S. S. McClure on the Left, and by intellectuals like Henry Adams on the Right, generation after generation of American novelists, Podhoretz writes, portrayed "their own society as corrupt, vulgar, philistine, materialistic, and puritanical."

What these novelists—the roster runs from Theodore Dreiser to Gore Vidal to last year's National Book Award winner, Jonathan Franzen—have rendered through invented characters and the power of the imagination has been amplified by the work of generations of social critics in the press and later in the universities. It was thus out of well-prepared soil that there would grow the "adversary culture" of the 1960s, a culture so persuaded of the unredeemable turpitude of the country sheltering it that it could liken that country to Nazi Germany.

Of course, by the time the sixties rolled around, the oppositional impulse Podhoretz describes had also been cultivated and fertilized by the critique of American capitalism, and of American democracy as its fake facade, that was the work of communist and left-wing radicals in the 1930s. That critique would be fortified during the cold war by the New Left's no less thoroughgoing assault on our supposedly racist politics at home and our supposedly regressive and "imperialistic" practices abroad.

But I am not just talking about the politics of a radical or revolutionary fringe. As contemporary historians have well documented, the ideas and opinions promulgated by the 1960s New Left and counterculture were echoed, in however diluted a form, throughout the institutions of the liberal mainstream, particularly the universities and the media. Nor was it only a matter of opposition to our involvement in Vietnam, or of outrage over lingering manifestations of racism at home. For a long time, cultural forces have been at work in the country that were not limited to questioning particular policies or to correcting particular inequities and injustices; they were aimed, instead, at the most fundamental arrangements and priorities of the American populace. When, in 1984, Jeane Kirkpatrick lambasted the "blame America first" faction within the Democratic party (the party to which she then still belonged), she was pointing to a disposition that by this time had moved close to the very center of American politics.

If one heard somewhat less about this disposition in the eighties and nineties, the age first of Reagan-Bush and then of Clinton, the attitudes behind it remained very much alive, consolidating themselves and waiting for an opportunity to reemerge politically. In those years, our literature, our popular music, and our arts became saturated in a nihilism that expressed itself in empty gestures of revolt or in celebrations of violence and sexual degradation. As for "one nation, . . . indivisible," racial and ethnic tensions were sharpened rather than ameliorated in the eighties and nineties by our official commitment to multiculturalism and "diversity." In government, education, industry, and every aspect of public life, an ethos of wounded resentment was fostered among minorities and majority alike. Nor did many of our professors and intellectuals abandon their reflexive suspicion of American aims in the world, and their no less reflexive sympathy for "liberationist" if not openly anti-American causes.

Much of the resultant tension appeared to come to the surface in the 2000 presidential election. For a moment, the country itself seemed to reflect—geographically, in those famous red and blue swaths of color on the map—the divisions that had been leading some observers to fear we were literally splitting apart, becoming, in the words of the distinguished historian Gertrude Himmelfarb, "one nation, two cultures." Could it be, others wondered in the aftermath of the Bush-Gore contest, that we were even on the verge of becoming two *nations*? "The two Americas apparent in the 48 percent to 48 percent election are two nations of different faiths," wrote the political analyst Michael Barone at the time. "One is observant, tradition-minded, moralistic. The other is unobservant, liberation-minded, relativistic." The rancor of the following weeks, as the outcome of the balloting hung in the balance and the fight over the presidency wound its way through the courts, only seemed to confirm this analysis.

The two Americas may have been equivalent in size, but they were far from equivalent in cultural influence or power. As Himmelfarb put it, the left-wing counterculture of the sixties and seventies—secular, liberationist, anti-traditionalist in every respect—had become dominant, while the culture of religious believers and moral traditionalists was in the position of a "dissident" element. However one reads the results of the 2000 election itself (and I'm well aware that some observers think the split vote reflected not the chasms in American society but, to the contrary, a movement toward the center), the point seems irrefutable. In terms not of numbers but of sheer cultural weight and prestige, at least until September 11, 2001, the flag-wavers among us represented a large but dissenting minority. The question is whether they still do.

THE WORKINGS OF LARGE CULTURAL movements are subtle, complex, and long-term. It takes a clarifying moment like the present one to expose their effects to the light of day, to reveal the territory that has been lost, and to bring home the effort that will be required before we can begin to think we might regain it. That lost territory is the territory of patriotism.

The problem is not that Americans are unpatriotic. That is hardly the case. The problem, to state it again, is that those who are *un*patriotic are, culturally, the most influential among us. To measure the success of the anti-American critique, one needs to monitor the attitudes not of those who espouse this critique in all its particulars but of those who may or may not accept any of its particulars and yet have been irrevocably affected by its underlying message. What was accomplished by the relentless critique of American reality and American ideals was this: It turned a simple and noble impulse, love of country, into a suspect category—or, just as corrosively, an unfashionable one. You did not have to approve of burning the flag to think that waving it was a breach of taste or manners.

Even today, when flags fly everywhere, these habits of mind and taste will be painfully slow to fade. Nor will they fade on their own. Things have moved very rapidly in the past months, but it was only yesterday that Barbara Kingsolver and Katha Pollitt were elucidating for their benighted fellow Americans what the flag *really* symbolizes. Just because sentiments like theirs have come to be stated less frequently, or less sharply, does not mean they have dissipated, or been answered. Still less does it mean that educators are busying themselves with instilling better ideas in the minds of their students.

An anecdote told by Chester E. Finn Jr. clarifies the issue. In the immediate aftermath of September 11, lunching with an "old acquaintance who now occupies a position of leadership in higher education," Finn found himself complaining about the materials that were already starting to come out to help teachers deal with the traumatic events. Those materials—I quoted from some of them in Chapter 2—were, Finn lamented, "long on pluralism but short on patriotism." Replied his companion: "I don't disagree with your point.

But do you have to use the word *patriotism*? It makes many of us uncomfortable."

This exchange helps to put into perspective quite a number of post–September 11 phenomena, including another incident I related in Chapter 2: the Pledge of Allegiance ceremony at Williams College, my alma mater, on the Sunday after the attacks. That event, organized by an undergraduate who was distressed by the implicit or explicit anti-American tone of faculty forums that had been held in the previous days, drew numerous students and maintenance workers, the college president, and exactly one professor.

Now, even at Williams, a very liberal school, I cannot believe that large numbers of faculty would say openly that we as a nation got what we deserved on September 11, or that we are as bad as if not worse than our attackers. "I don't disagree with your point," some Williams professors might even say, if pressed on the issue of fighting back and punishing those who hit us. But actually coming out, in public, as partisans of America? That, clearly, would make them "uncomfortable."

Another example: Back at the height of the Vietnam War, the faculty of Harvard University, where I earned my law degree, decided to join the protest against America's involvement in Southeast Asia by withdrawing all curricular and academic status from the Reserve Officers Training Corps (ROTC). The Vietnam War ended in 1975, but today, over a quarter-century later, the ban on ROTC at Harvard still stands, justified now by disapproval of the military's policy on homosexuals. The ban is steadfastly defended by a faculty that may regret the hypocrisy involved—although there are ROTC students at Harvard, who attend college with the aid of federal scholarship money, they are compelled to take their training elsewhere, out of sight—but evidently not enough to make an open and honest commitment to national military service. The university even refuses to grant credit for ROTC courses taken at other schools.

The hypocrisy is hydra-headed. While closing the door to ROTC, Harvard accepts money from the wealthy and powerful bin Laden family of Saudi Arabia for programs in Islamic law, art, and architecture—the same family whose infamous scion stood behind the attack on our soil. Exercised by our military's "don't ask, don't tell" policy toward homosexuals, Harvard also seems altogether untroubled by the policy of the Saudi regime in this regard: On January 1, 2002, the kingdom announced its first three beheadings of the year for homosexual conduct. What do these facts tell us about where the university chooses to draw its moral lines?

But more is at stake here than hypocrisy. Harvard's action, Ruth Wisse has written, violates the most basic principle of democratic citizenship. Its implication is that any group of Americans "who are dissatisfied with one or another government policy need not be prepared to protect the country as a whole because it does not live up to *their* concept of the good." Wisse is herself a professor at Harvard, where her arguments go unappreciated by most of her fellow faculty members but are shared by more and more students. Pressure from these students, as well as from a number of outraged alumni, may yet result in ROTC's reinstatement.

But the task of making our educators "comfortable," as Americans, looms as large as ever. For that to happen, at Harvard, Williams, or elsewhere, there will have to be a deep revolution in consciousness among a class of people trained for generations to believe that being an intellectual and being a patriot are two wholly incompatible callings.

THE FIRST OBSTACLE TO BE overcome is sheer, unadorned ignorance. For "what is not taught will be forgotten, and what is forgotten cannot be defended."

In these pithy, precise words, the American Council of Trustees and Alumni recently summarized the utter failure of our institutions of higher learning to provide college students with a thorough grounding in the history and civilization of their own country. In a survey released in late November, the council revealed that of the fifty-five top-ranked universities in the nation, not a single one required a course in American history, and only three required a course in Western civilization. No wonder, then, that in an earlier Roper survey of graduating seniors from these same top schools, more students named Ulysses S. Grant than George Washington as the commander of American forces at Yorktown, and only 22 percent knew that the phrase "government of the people, by the people, and for the people" came from the Gettysburg Address.

The ignorance of our college students is old news; even older news is the ignorance of students, and teachers, at lower levels. That ignorance has been documented over and over again in international comparative surveys in which American children regularly score at or near the bottom in the mastery of everything from math to history. It is the latter subject that concerns me here. If our students do not know their nation's past, have never been informed about the ideas and values of their society, how are they ever to participate knowledgeably in our national life? If they never learn about the rest of the world, how are they to understand the challenges to American principles and values, or what it has taken to defend and protect them, over the last two centuries and still today?

This is the vacuum into which have stepped the diversity-mongers, the multiculturalists, the relativists, and the plain old anti-Americans. It is a vacuum created in part (as Walter Berns shows in his book *Making Patriots*) by a series of Supreme Court decisions dating back to 1947, invalidating laws whose purpose was "to endow children with a love of country, and to transmit this from one generation to the next." It is sustained by a cultural and moral relativism that, in place of teaching students to love their country and be prepared to make sacrifices for it, overlays their abysmal ignorance of its history with a "sophisticated" understanding of America as but one cultural option among many of equal worth, and then replaces the impulse to love and defend it with detachment, indifference, or shame.

Once we could depend on our elementary and secondary schools to do the basic job of both imparting knowledge and nurturing civic pride. But the very idea of what used to be called "civics" is now anathema, evoking as it does an image of unity and coherence in an age whose pedagogical mantra is that we are instead a collection of races and ethnic groups only accidentally thrown together and in need above all of protecting our separate identities. And besides, what is there to be proud of? Many of our children, as I said earlier, are taught from the earliest grades that there are no differences among cultures and that ours deserves no preference. They're lucky; older children learn that Americans have much to apologize for, having stolen the land from its native inhabitants, despoiled the environment, enslaved an entire population, made off with territory belonging to Mexico, mistreated women, exploited the laboring classes for the benefit of robber barons, discriminated against immigrants and people of color, and wantonly sent young men to die in imperialist wars against the defenseless and poor of the third world. At the college level, one looks mostly in vain for correctives to this teaching.

Now we need to defend what has been either forgotten or violently misrepresented, not only on real battlefields with real weapons but on the battlefield of ideas with words and facts, with concepts and arguments. And now, too, we can begin to appreciate the price of our children's ignorance.

Not that any of the proponents of today's insidious pedagogy would admit what they have been up to. "If our aim is to indoctrinate students with unpatriotic beliefs," said the leftist historian Eric Foner of Columbia University in late November 2001, "we're obviously doing a very poor job of it." Foner was referring, sarcastically, to a poll disclosing the surprising news that there was firm support for the war among college students nationwide. His comment was intended to ridicule the notion that professors like him do strive to influence their students' views against their country, but it also inadvertently conveyed the sneering mind-set against which those students, like the students at Harvard fighting for the return of ROTC, are going to have to struggle. His very choice of words—"indoctrinate," "unpatriotic"— seemed calculated to caricature and impugn the presump- tively McCarthyite motives of anyone who would dare ques- tion Foner's own motives, let alone his thoughts.

A defensive tactic, in short, and perhaps understandable enough in the circumstances. For this much-honored author, who has made his reputation as a left-wing basher of America, is a capitalist to this extent: He has a big investment in the marketability of his views, and an even bigger one in the still-malleable loyalties of his students. To be confronted with an upsurge of patriotism in his own bailiwick must be threatening indeed.

Of course, I do not mean to single out one professor—I am speaking of a class—but this does happen to be the same Eric Foner who could bring himself to assert, even while the fires were still burning in late September, that "I'm not sure which is more frightening: the horror that engulfed New York City or the apocalyptic rhetoric emanating daily from the White House." What atrocious rot. And yet such pro- nouncements, and worse, once tripped lightly and almost without the necessity of thought from the mouths of people like him. It was all part of the culture wars, and it cost them nothing to say whatever wild and reckless things they pleased about their country and its leaders—on the contrary, they were richly rewarded for it, and they were winning.

But now thousands of their innocent countrymen were ash, and there was a smoking hole in downtown New York, and another in the side of the Pentagon, and a third in a field in Pennsylvania, and anyone with eyes to see and ears to hear could tell the difference between one thing and another, between the horror that had engulfed New York City and the determined, manly, but far from apocalyptic rhetoric of the president. Foner could not have been alone among the professoriate in thinking it might be time to try a new tack, time to pretend "we" had never tried to indoctrinate anybody—and by the way, we failed.

ARE THE FONERS OF THE UNITED States in fact doing "a very poor job" of indoctrinating students with anti-Americanism? Polls have seemed to suggest they are, and so does much anecdotal evidence of the kind I have presented here. That is a very good sign. But it is not enough for the Foners to fail. Others have to succeed. To put it another way: It is not enough (though it certainly helps) for our students to have the right instincts. They have to have knowledge, too.

A vast relearning has to take place. The burden of this relearning falls upon all of us—public officials and private individuals, clergymen, politicians, military personnel, civilian authorities of every kind, mothers and fathers. But most especially it falls on educators, and at every level. The defect can only be redressed by the reinstatement of a thorough and honest study of our history, undistorted by the lens of political correctness and pseudosophisticated relativism. This is not jingoism; it is a call to repudiate the mind-set that has encased the teaching of our history in relativist and anti-American myth and to replace it with a genuine inquiry into fact and a genuine openness to debate. I, for one, am hardly in doubt as to the outcome.

We learn history, said the philosopher Leszek Kolakowski of the University of Chicago in his Jefferson Lecture in 1986, "to know who we are," to learn "why, and for what [we are] responsible," and to acquire a "historically defined sense of belonging." This is especially important in the United States, a nation created to realize a specific political vision. For it is our collective memory of that vision—the American idea—that defines us as Americans and ineluctably exerts its pull on our patriotic emotions. (Lincoln in his soaring language referred to this when he invoked the "mystic chords of memory" that, "stretching from every battlefield and patriot grave to every living heart . . . swell into a mighty chorus of remembrance, gratitude, and rededication.") By studying our history, by learning about its heroes, by examining and understanding its failures as well as its incomparable achievements, we grasp the value of our political tradition and what distinguishes it from others.

Our country *is* something to be proud of, something to celebrate. Why should we shrink from saying so? A sober, a sophisticated, study of our history demonstrates beyond cavil that we have provided more freedom to more people than any nation in the history of mankind; that we have provided a greater degree of equality to more people than any nation in the history of mankind; that we have created more prosperity, and spread it more widely, than any nation in the history of mankind; that we have brought more justice to more people than any nation in the history of mankind; that our open, tolerant, prosperous, peaceable society is the marvel and envy of the ages.

This is demonstrably true within our borders. And outside our borders? We have been a beacon of freedom and opportunity to people throughout the world since the day of our creation. America is the place people run *to* when, in hope or hopelessness, they are running from somewhere else. As for our record of alleged "imperialism," so richly documented by our nation's critics, an English columnist has answered all such charges by inviting us to

> ponder exactly what the Americans *did* in that most awful of centuries, the 20th. They saved Europe from barbarism in two world wars. After the second world war they rebuilt the Continent from the ashes. They confronted and peacefully defeated Soviet Communism, the most murderous system ever devised by man. . . . America, primarily, ejected Iraq from Kuwait and . . . stopped the slaughter in the Balkans while the Europeans dithered.

This list could be extended tenfold, and still the story would be incomplete.

A positive assessment of American history is not the same thing as an uncritical assessment. If we were created by a political vision, our story is the story of a struggle to realize that vision. A struggle has its ups and its downs, its advances and setbacks; it is subject not only to changing circumstance and to the shifting quality of leadership in any generation but to the vicissitudes of human character and the enduring waywardness of the human heart. We have certainly had our failures, some of them shameful. But never once, I think, have we lost sight of our moral ideals, which is why, time and again, we have succeeded in confronting, overcoming, and transcending the stains on our record, the stain of slavery foremost among them. Who among the nations can enter a similar claim?

"Am I embarrassed to speak for a less than perfect democracy?" asked New York's former senator Daniel P. Moynihan, who, as our ambassador to the United Nations in the mid-1970s, withstood with force and flair the seemingly irresistible ideological assault within that institution on Western and American values. His answer:

> Not one bit. Find me a better one. Do I suppose there are societies which are free of sin? No, I don't. Do I think ours is, on balance, incomparably the most hopeful set of human relations the world has? Yes, I do.

It is starting from here, from this bedrock understanding, that an education in patriotism should proceed. "In telling the story of the American political experience," the Yale historian Donald Kagan has said, "we must insist on the honest search for truth." For patriotism "does not require us to hate, contemn, denigrate, or attack any other country, nor does it require us to admire our own uncritically." What it does require, what we have systematically denied our young, is a common educational effort to absorb and to transmit "this country's vision of a free, democratic republic and its struggle to achieve" that vision.

As for those who shirk or abjure this effort, like the voices we have encountered in these pages, Kagan condemns them in tones ringing with indignation and truth. Such assaults on patriotism, he writes, are "failures of *character* [my emphasis], made by privileged people who enjoy the full benefits offered by the country they deride and detest, its opportunities, its freedom, its riches, but [who lack] the basic decency to pay it the allegiance and respect that honor demands." This, in turn, imposes a different, double responsibility on the rest of us who think, and know, otherwise—the responsibility to point out "the despicable nature of [the critics'] behavior" while at the same time respecting their right to be irresponsible and even subversive of our safety. "Our own honor, and our devotion to our nation's special virtues," requires nothing less.

There is much more of golden value in this text by Kagan. He delivered it as a public lecture at Yale, in reply to what he deemed the dishonorable response of a number of his fellow professors to the events of September 11. His talk, which drew a standing-room-only crowd, elicited a sustained, clamorous ovation from students starved for confirmation who had come to hear that their deepest feelings about their country were not, as they had been instructed, unworthy, but on the contrary the very core of civic sense. How utterly liberating to be told, by one of the foremost scholars in the land, that they possessed within their already mobilized emotions the basis of an education in the "civic devotion and love of country we so badly need." I hope that the Yale student I cited in Chapter 2, who lamented her moral miseducation, was in the audience.

CIVIC DEVOTION, LOVE OF COUNTRY. From these there flows national unity, that precious commodity that seemed to erupt with such spontaneous immediacy after September 11 but that otherwise has seemed, over the last decades, in such desperately short supply. "We are an enormously diverse and varied people," Kagan writes, "almost all immigrants or the descendants of immigrants." Although this diversity has given us tremendous strength, it is not without its dangers, primarily in the form of divisions that set one group against another. And insofar as these divisions prevent the forging of a single people, they also prevent the building of a true and thoughtful patriotism. In our time they have been permitted, or actually encouraged, to fester and grow; in the absence of an offsetting education in the American national idea, they have fostered a debilitating "reluctance," in Kagan's words, "to work toward the common good and to defend our country when defense is needed."

This is the time when defense is needed, this is the time for legitimate patriots. No one who has read the newspapers or watched television over the last months can fail to have been thrilled by the response of ordinary Americans to the hour of emergency. But today's ordinary Americans live off the stored-up moral capital and the living memory of another century's experience.

The schools of that century are not our schools, the experience of that century is not our experience, the values of that century . . . are they still our values? How can we expect our children to meet tomorrow's hour of emergency as we would wish them to if we neglect to instruct them in civic devotion, and love of country, and in the certitude that the United States is one nation, indivisible?

We must not squander the opportunity.

WHY DO
WE FIGHT?

HERE ARE THE WORDS OF Osama bin Laden, in 1998: "By God's leave, we call on every Muslim who believes in God and hopes for reward to obey God's command to kill the Americans and plunder their possessions wherever he finds them and whenever he can." And: "To kill Americans and their allies, both civil and military, is an individual duty of every Muslim who is able."

Here is bin Laden again, after the attack on September 11: "God has blessed a group of vanguard Muslims, the forefront of Islam, to destroy America. May God bless them and allot them a supreme place in heaven."

In the ruins of Kabul in late December, an enterprising *Wall Street Journal* reporter came into possession of abandoned al-Qaeda computers. On the hard drives were found memoranda detailing plans for manufacturing and assembling the materials for large-scale biological and chemical warfare against the West—so much cheaper and easier, an al-Qaeda commander had noted in the computer file, than getting hold of nuclear weapons, and every bit as effective.

That is why we fight.

"GIVE ME A BREAK, I'M HUMAN."

John Burke, assistant headmaster at Xavier High School, a Jesuit school in New York, is the father of a retarded child and has a grandson who is blind; another son, Xavier class of '91 and a broker at the firm of Cantor Fitzgerald, died on September 11 in the attack on the World Trade Center. So did nine other alumni of the school; so did the fathers of three boys; so did dozens of other relatives of students, teachers, and staff members. There were sixty dead in all. Since September 11, teachers and administrative personnel had been dealing with their own private grief, the grief of those students most closely affected, the grief of an entire school of one thousand that had to face not only devastating personal loss but the starkest questions that can ever confront a community of faith. How could a loving God permit such terrible, unmerited suffering?

But here is the miraculous thing. Precisely because it is a community of faith, Xavier is a lucky place, a place you would want to be a part of at such a moment of supreme crisis.

Both elements, the community and the faith, are crucial.

Community: "[E]veryone who has ever gone to Xavier knows everybody else, or so it seems," wrote a clearly awestruck *New York Times* reporter. "They played on the same Catholic Youth Organization basketball teams, went to the same colleges, surfed the same beaches, worked at the same firehouses and brokerage firms." Faith: It is a place that lives by ritual, by order, by prayer, by a tradition that asks ultimate questions and is prepared to grapple with the answers, however incomplete, that have been offered over thousands of years by some of the subtlest minds ever created.

"Of course it tests your faith," said John Burke about September 11 and the tragedy it injected into his community. "But faith is a choice, and this is an environment that nurtures it."

I am a Catholic. I am familiar with these communities, with their unique ways of nurturing and sustaining those within them. I also have no illusions about them; I recognize that they are not for everyone, and why some have fled them, and communities like them, for unfettered lives as well as for lives with fetters of a different sort. Such unfettered lives, indeed, are precisely what increasingly large numbers of secular (and formerly religious) Americans have chosen for themselves over the last decades, in some cases defending their choice as a more responsible, a more mature, path to human happiness. (Others admit that it is an easier path.) But to put the matter at its simplest, isn't there something that these institutions, in their accumulated wisdom, *know* about the souls of mortals, especially mortals who are badly hurting, that is simply unavailable anywhere else in our society, and that we ignore at our collective peril? And is there not a surpassing value in what they have to impart to us about the reciprocating qualities of faith, hope, and charity, a message at which we should never scoff but which we should instead honor and be at pains to recover?

At Xavier, when September 11 came, followed by its long, tormented aftermath, a little platoon was already in place to keep watch over the traumatized boys, to comfort and guide them through their unspeakable grief, to relearn, with them and even from them, the stubborn, insistent necessity of affirming life. "I was really down," the Reverend John Garvey, who teaches religion at Xavier, told the *Times* reporter. "I had the 'whys.'" But then, he continued, a student whose father, a firefighter, had died in the conflagration came to see him. The boy had recently had an encounter with the members of his father's engine company, and they had filled him with immense pride in who he was and where he came from, and this pride then communicated itself to his despairing teacher.

"My presumption was I had to take care of this boy, swallow my grief and be there for him. But he reminded me," said Father Garvey, in wonderment and gratitude, "that we had to go, be, do, act, stop the 'what ifs' and deal with what is. He was the voice of hope, love, and caring."

That is why we fight.

TODD BEAMER, JEREMY GLICK, and Tom Burnett were passengers on United Airlines Flight 93, hijacked by al-Qaeda terrorists on September 11. Like the other doomed planes that had been seized by terrorists that morning, this one was intended to be used as a missile, probably to destroy the White House or the Capitol in Washington. But Beamer, Glick, Burnett, and others thwarted the terrorists' intention, forcing the plane down in rural Pennsylvania—at the cost of their own and their fellow passengers' lives, but saving the innocent lives of countless others and preventing still another devastating blow to their country's national symbols. *They* became, instead, national symbols themselves.

Who were these men? After September 11, we were regaled with stories of the firefighters and policemen and rescue workers who had charged into the cauldron in downtown New York, trudging with unflappable manly determination up the smoke-filled stairways of the World Trade Center, into certain death. To this day I cannot read about them without being moved by their stoic courage, the fierce bonds of brotherhood among them, the sense of duty so ingrained in their character as to be not second but first nature. Said the widow of one of them who did not return, the father of ten: "If that had been a New York mosque burning and there were Taliban inside, he and his brothers would have gone in."

"I know what you all did," Captain James Gormley addressed the comrades who had gathered on December 10 at the last funeral service for the thirteen men of a single engine company on Manhattan's West Side who perished on that awful day. "You got your gear on, found a tool, wrote your name or Social Security number in felt-tip pen on your arm or a leg, a crisis tattoo in case you got found." Captain

Gormley went on:

> We went down there knowing things could go badly.
> We stayed until we were exhausted, got three hours'
> sleep and went back again, and again. That's what
> comrades do. Only luck and circumstance separate
> us from them [the dead].

These were our heroes. I take nothing away from their heroism when I note that they were also trained to do what they did, that it was their job. But who were Beamer, Glick, Burnett, and the rest? They were guys in jackets and ties, guys in white shirts, businessmen, family men, representatives of the great American middle class, the most maligned class in history.

Beamer was a business executive from suburban New Jersey, married, the father of two boys ages four and two, with a third child on the way. Glick was an Internet salesman from Boston, married, the father of a twelve-week-old daughter. Burnett was a vice president at a medical-research company, married with three children. Beamer was a stalwart of his church, deeply committed to his faith; Glick was Jewish though not particularly religious. All of them had been high school or college athletes. Three more typical men would be hard to imagine.

But typical of what? Only a couple of years ago, the movie *American Beauty* had set out to capture the spirit of American suburbia today, the land of middle-class businessmen like Beamer and Glick and Burnett: married, white, comfortable . . . and utterly lost. The main character in the movie, Lester, is played by the actor Kevin Spacey, who had this to say about his role: "I think Lester is very much like a lot of men in American life, who start out having certain ideas about the kind of life they will have, but somewhere along the line, they get squashed down." Lester's wife is named Carolyn. Annette Bening, who plays her, commented: "Carolyn is like so many people who feel an emptiness in their lives and try to fill it up with having the right things. She believes if you have the right car and the right house and even the right garden, then somehow your life will turn out all right." As the movie unfolds, these two moral ciphers and their various neighbors and kin, presented to us as quintessential products of our culture of material success, proceed to wreak havoc on themselves and everybody around them as Lester undergoes what the filmmaker seems pleased to regard as a cathartic liberation, if not a spiritual rebirth, from the American condition.

What lesson, exactly, do we learn from *American Beauty*, if not the lesson preached incessantly by our cultural elites over the decades: that America is itself a kind of death? And what lesson do we learn from the counterexample of Todd Beamer and Jeremy Glick and Tom Burnett? I can answer that one easily enough: Three months after September 11, inspired by their act, passengers and flight attendants on an American Airlines flight sprang into action to overpower a man trying to set off a bomb in his shoe that would have ripped the plane from the sky. Nihilism is contagious, and is truly a kind of death; citizenship and mutual responsibility are also contagious, but they lead into life. And, oh yes, *American Beauty* is a lie.

On the morning of September 11, Jeremy Glick furtively telephoned his wife from onboard the hijacked United Airlines flight to confirm the rumor that in the last few minutes other planes had already crashed into the World Trade Center. When she told him it was true, he passed the information on to Beamer and the others. "We've decided," he reported back to his wife after the men had conferred; "we're going to do it." Tom Burnett had also consulted with his wife. "I know we're all going to die," he said to her. But "there's three of us who are going to do something about it." Todd Beamer had a telephone line open to a GTE operator in Chicago. He made her promise to contact his wife, then she heard him recite the Lord's Prayer, and then came the defining words: "Are you guys ready? Let's roll."

That is why we fight.

UNDER THE TALIBAN, MUSIC was banned in Afghanistan—all music, except for unaccompanied religious songs and chants. As a British musicologist recounted at the time, "All musical instruments in the possession of Afghan performers have been confiscated and burned. Those caught in possession of musical instruments are imprisoned, fined, or even beaten, and their instruments are destroyed." In one characteristic incident, goons from the Taliban Ministry for the Promotion of Virtue and the Prevention of Vice paid a visit to a seller of tape cassettes in the bazaar in Kabul; they pulled the tape out of all his cassettes and hung it from the trees.

Not obscene or lascivious music; not violent music; not

ugly music; not even all Western music, the music of the "infidels"—with the one exception noted, music of every kind fell under proscription. The Taliban's ban on music was of a piece with its enforced wearing of the burqa, the head-to-toe covering for females, or the enforced wearing of regulation-length beards by men. It was of a piece with the closure of theaters, with the beating or imprisonment meted out to anyone caught playing chess, with the proscription of sports clubs. It was of a piece with the Taliban's main idea of positive entertainment, which was the public execution of adulterers. But the ban on music also had a special meaning of its own, for, like no other art, indeed like no other human activity, music speaks directly to the soul, awakening longings of transcendence and bearing intimations of other, better, wider, more light-filled worlds. In banning it, the Taliban knew well what they were doing.

When Kabul was liberated in mid-November, the public scaffolds were torn down. Kites, which had also been banned, flew gaily in the streets. Men crowded barbershops to have their beards shaved off. Women slowly began to show themselves in public, to resume their studies and their careers. And everywhere, everywhere, there was singing, and the sound of music. In the bazaar, the seller of cassettes loudly played a recording of the great Afghan singer Ahmad Zaheer: "Today I have drunk too much, let me dream my dream."

The music was playing, and people were dreaming their dreams again.

That is why we fight.

NOT LONG AFTER THE TURN of the twelfth century, a young genius by the name of Peter Abelard arrived in Paris. He would be known forever in the annals of love for his tragic romance with a young woman named Héloïse, but in the course of his turbulent lifetime he became famous for other things as well, including his many disputes with his masters, his groundbreaking treatises in logic and philosophy, his directorship of the cathedral school in Paris, his foundational works in Christian theology.

Abelard introduced into theology the *sic et non*—yes and no—method that was then current in the discipline of law. This was a mode of logical study that proceeded by dialectical reasoning—the weighing of opposite propositions in an open-ended search for the truth. In the following centuries, it would become the organizing principle for the academic disciplines of both philosophy and theology, if not the basis of all learning itself. In the broad sense, *sic et non* might be thought of, indeed, as a watchword of our Western civilization, of our very outlook on the world.

Is Western thought, as its many critics contend, the product of dead, white, European males? To a large extent—but by no means wholly—it is. Is it therefore homogeneous, monolithic, closed? The charge is as absurd as it is spurious. To ascertain the value of capitalism, we consult Adam Smith on the one hand, Karl Marx on the other. To weigh the value of the religious life, we go to both Aquinas *and* Voltaire. On the utility of warfare, we read Homer *and* Erasmus. For wisdom on the ends of politics, we study Aristotle *and* Machiavelli. We do not, like the relativists, suppose that there is no such thing as the truth; we do posit contention in getting to it.

In railing against the "self-righteous drivel" that what we had endured on September 11 was an "attack on civilization," the intellectual superstar Susan Sontag wanted to focus instead on politics, and in particular on the culpability of "specific American alliances and actions." But she was wrong about September 11, both specifically and generally, and she was also denying the wellspring of the civilization that sustains her. For the open, curious, free spirit of *sic et non* is what differentiates us from totalitarianisms of every kind, political and religious alike, and is one of the main reasons they hate and seek to destroy us. It is our glory as a Western culture; it stands behind our evolving institutions, and behind our unchanging values; it is the essence and the engine of civilization.

And that, too, is why we fight.

"I RECENTLY TOLD AN ASSEMBLY at my son's high school that they were living through a time so blessed they would tell their grandchildren about it," wrote the columnist Charles Krauthammer at the end of 1997. "They looked at me uncomprehendingly, first, because they have known little else but good times; and second, because it is hard for anyone to apprehend and appreciate the sheer and total felicity of the time in which we live until it is gone."

It is true: We were living in a bubble in the 1990s, and as long as we were inside that bubble, we as a nation permitted

ourselves every sort of indulgence, moral as well as material.
For who could imagine that bad news would come, or what
shape it would take, or what it might demand of us? In those
days, the most frightening prospect we could conjure up, as
Krauthammer facetiously put it, was El Niño, or maybe three
shark attacks. But a bubble it was, and eventually it was sure
to be revealed as one. "Golden ages do not last,"
Krauthammer warned. "There might be a sudden crisis, a
catastrophe, a collapse . . . or just a gradual undoing of all
this self-reinforcing good news." But however it would happen, this "lovely world" of ours would surely end.

He was right; it did end. The bubble burst on September 11, when the unimaginable became real and suddenly we
found ourselves in an entirely new and entirely unlovely
world: vulnerable, warred upon, and at war. The enemy—
who, or what, was it? Our government—where had it been
all those years? The outcome—what would it look like? The
military campaign—could we actually win it? The home
front—would we be required to transform ourselves, and
especially our liberties, beyond recognition? Our national
spirit—was it up to the challenge?

In the weeks and months to follow, we would learn some
of the answers to some of these questions, and those answers
would reassure us. The first phases of the military campaign
in particular went spectacularly well, bringing about the swift
collapse of the Taliban regime in Afghanistan and the defeat
and scattering of al-Qaeda forces from that country. By year's
end, the terrible deficiencies in our past intelligence-gathering and, more important, in our foreign policy were being

addressed and slowly overcome. Our economy, still shaky, seemed poised for recovery in the foreseeable term. Our liberties, despite dire and even frantic warnings to the contrary, were secure.

But what about our spirit? To me, that precious and indefinable quality, whose health has been my abiding preoccupation over the last decades, remained, and remains, a cause of worry. A few days after the September attacks I had kept a scheduled commitment to speak at a fine school, Tyler Junior College, in Texas. There I said that, though we had just become, through the deaths of our fellow citizens, a radically diminished country, in some ways we were also a better country: focused on, and thinking about, more important things than had been the case before September 11. At a terrible cost, we had been reminded of what mattered, we had had our perspective restored; this was a necessary and a salutary thing.

Still, I said in mid-September, I was concerned about the future. I was afraid that, as a country, we had had it too good during the years of the bubble; having been softened up, we might not be able to sustain collective momentum in what we were now being called upon to do. To be sure, I said, I had every confidence that our president, and our military commanders, would see us through to the end of the larger war on terror to which the president had committed us. My real worry was whether the American people themselves would persevere, whether they would rise to the full implications of this moment of moral clarity, turn the coming age into an age of moral clarity, and not permit their own magnificent response to September 11, or the magnificent selves they had revealed in those days, to fade into just another distant memory.

Today I still have confidence in our political and military leadership, more confidence than ever, though I am also well aware that, precisely because of our rapid victory in Afghanistan, pressures can only mount to quit while we're ahead, to call off the much more arduous campaign still awaiting us. What I do continue to worry about, especially as we go forward, is how to keep up our spirit. To say it for the umpteenth time, I do not fear for our instincts; by every indication those are as sound as they disclosed themselves to be in the immediate aftermath of September 11. What I fear is the erosion of moral clarity, and the spread of indifference and confusion, as a thousand voices discourse with energy and zeal on the questionable nature, if not the outright illegitimacy, of our methods or our cause.

We have been introduced to a sampling of such voices in the course of this book. For a concrete instance of the turmoil they can introduce into public discourse, take the frenzied debate over civil liberties to which I alluded a moment ago, and in particular over the issue of military tribunals for foreign terrorists—"secret" military tribunals as they were often characterized by the press, though no one had said anything about secrecy. Even as the Pentagon was still writing the rules for these tribunals, we were being warned loudly and with perfect assurance that they would inevitably turn into kangaroo courts, without protections of any kind for defendants. "It sounds unconstitutional," fretted the

Washington Post columnist Mary McGrory; "our European allies are appalled." By the time the details of the Pentagon's rules emerged in late December, the poison of suspicion had already entered into the cultural bloodstream.

As the war winds on, especially if the fighting becomes more difficult and if patriot graves should, God forbid, multiply, such voices will only strengthen. The military battle is one thing. The battle of public opinion, over our airwaves and in our newspapers and journals, in our schools and churches, in our families, in our hearts, is another. We have to understand that not only our strength of arms but our character is being tested, and so is our mettle, our staying power. The temptation will be great to call it a day while we are still in night—disregarding what lies in wait for us if we should falter, belittling how very much depends on us, demeaning the incomparable blessing that will be ours, and our posterity's, if we prevail.

We cannot allow this. We can never allow ourselves to forget why it is that we fight; why we *must* fight.

IRAQ

ON THE HEELS OF PRESIDENT BUSH'S speech to the United Nations on September 12, 2002, a year and a day after the terrorist attacks on the United States, the Security Council passed Resolution 1441. This unanimous resolution stated, among other things, that Iraq was in breach of over a dozen previous U.N. resolutions going back to its invasion of Kuwait in 1990. Now Iraq would be given "a final opportunity to comply with its disarmament obligations."

President Bush hardly needed a U.N. resolution to vindicate the longstanding position of the United States that Saddam Hussein was armed and dangerous, and should be removed from power. The evidence was and is plain to the human eye, and the imperative for American action had been codified by the U.S. Congress, which expressed its own resolve on the matter in the Iraq Liberation Act, passed during the Clinton administration. That resolve was expressed again in October 2002 when the House and Senate passed a joint resolution authorizing the president to "use the Armed Forces of the United States" to defend our national security "against the continuing threat posed by Iraq . . . and to enforce all relevant United Nations Security Council resolutions."

The reason President Bush went to the U.N. was that he wanted to bring the civilized nations of the globe into the fold of the American war against terrorism. He was also aiming to retrieve the rather badly tarnished reputation of the U.N. itself and, along with it, international norms of civility. Receiving the backing of the U.N. would thus be a kind of test: for years, that body had not only failed to act against the tyrant of Baghdad but had become prey to a gaggle of anti-American authoritarian regimes. Would it now stand at last against barbarism, and in solidarity with the world's greatest democracy? President Bush, in the end, went to the U.N. to defend *its* honor, not ours.

In issuing Security Council Resolution 1441, the U.N. did indeed seem to pass a test of common sense, which is the true test of our times. Recapitulating the history of Saddam Hussein's repeated violations of its own dictates, it provided him one last chance to tell the truth about his arms programs.

The outcome of this initiative may or may not be known by the time this book is published. But common sense should have left little doubt that, whatever subterfuges he might engage in, Saddam Hussein would never disarm. In the words of Khidhir Hamza, the former head of Iraq's nuclear weapons program, "Saddam will never come clean." Nor could one count on the zealousness of the U.N. inspectors who were soon dispatched to verify Iraq's compliance with the terms of the U.N. resolution. Nor, finally, even if incontrovertible evidence of continuing weapons programs were to be uncovered, could one confidently anticipate that the U.N. Security Council would call for international force in order to bring about a change of regime.

And yet, who could doubt that, at this late date, nothing less than armed force would do if the world were to be rid at last of the scourge of Saddam Hussein?

Well, it turns out that many could and did doubt that very thing. By the summer of 2002, the idea that Iraq might indeed become a target of American action had aroused the passions of the antiwar brigade. In November, notwithstanding even the unanimous vote of the Security Council in passing Resolution 1441, fresh protests cropped up all across the country. Hollywood personalities from Barbra Streisand to Susan Sarandon to Ed Asner publicly challenged President Bush on the "need" to oust Saddam Hussein. Students and professors on the nation's campuses found a new cause of the moment. Antiwar petitions flooded the Internet, and the newspapers carried full-page advertisements signed by political activists, "concerned" business leaders, churchmen, academics, and human-rights advocates.

Even current and former elected officials raised their voices. In September 2002, Democratic congressmen Jim

McDermott of Washington and David Bonior of Michigan traveled on a fact-finding mission to Iraq. Interviewed on ABC's *This Week*, McDermott was asked if the Iraqi regime could be trusted to allow unfettered inspections. From Baghdad, McDermott said: "I think you have to take the Iraqis on their face value." Asked a little later in the interview if he believed President Bush to be a liar, McDermott replied: "I think the President would mislead the American people." To an elected member of the U.S. House of Representatives, speaking on enemy soil, the butcher of Baghdad was to be taken at face value, the word of the president of the United States to be held under suspicion.

McDermott was not the only representative of the Democratic Party to condemn the president. When, earlier in the year, Bush spoke of an "axis of evil" comprising North Korea, Iran, and Iraq, former president Jimmy Carter called this an "overly simplistic and counterproductive" statement that would take "years to repair." This was the same Carter who, as president, had watched helplessly as Iran fell to the Ayatollah Khomeini, as fifty-two American hostages were subsequently seized and held in Teheran, as Nicaragua fell to the Soviet-supported Sandinistas, and as Afghanistan fell to the Soviet Union—and who is now the recipient, incredibly enough, of the Nobel Prize for Peace. Adding insult to ignominy, Carter received his award the same year it was discovered that North Korea had become a nuclear power. Carter's mission to North Korea, during the Clinton administration, had helped to facilitate that dangerous dictatorship's own program of weapons of mass destruction.

And then there was former vice president Al Gore, who in a well-publicized speech at the Commonwealth Club in San Francisco assailed President Bush for concentrating on Iraq at the expense, presumably, of the war on al-Qaeda. This effort, said Gore to loud applause, was "unfortunate," squandering the "enormous reservoir of good will and sympathy and shared resolve" that we had built up and replacing it "with great anxiety all around the world, not primarily about what the terrorist networks are going to do, but about what we're going to do." In Gore's analysis, then, the world had more to worry about from U.S. actions than from terrorist actions—more to fear from those who would deliver the oppressed from evil than from the oppressors and the doers of evil, more to fear from a peace-loving democracy than from a ravening dictatorship.

FROM LISTENING TO VOICES like these, one might have thought there were grounds for doubting the strength of the case against Saddam Hussein—that he was not the wicked tyrant he had been pictured as being. Nothing, however, could be further from the truth.

In the war that the Iraqi dictator launched against Iran in 1980, and that would continue for eight long years, he killed more than a half-million Iranians, with many estimates running much higher than that. During this war, and in violation of the Geneva accords concerning the use of chemical weapons, Saddam Hussein deployed both mustard and nerve gas. In 1988, he unleashed chemical weapons against the civilian population in Halabja, killing at least five thousand people (some observers triple and quadruple that number). In 1990, he invaded and took over the sovereign country of Kuwait. By 1991, he had killed more Muslims than any other person in modern history.

A lot more than this was known, and on the public record, for all to see long before November 2002. It was known that Iraq had violated more than a dozen United Nations resolutions relating to weapons of mass destruction. On November 5, 1998, the U.N. Security Council in Resolution 1205 found Iraq in "flagrant violation" of the prior demand of Resolution 687 that Iraq destroy such weapons and unconditionally agree not to acquire or develop new ones. Military action against Iraq was to be sanctioned if it failed to comply with the terms of Resolution 687—and yet Iraq remained in "flagrant violation," and remained unpunished.

Also known was that the Clinton administration had supported regime change in Iraq. In 1998, President Clinton said that if we allowed the situation there to fester, Saddam Hussein would inevitably conclude that he was free to rebuild his arsenal of mass murder. "Some day, some way, I guarantee you," said the president, "he'll use [that] arsenal."

Known, too, was that in 1991 Iraq had categorically denied having a biological-weapons program. But in 1995, in the wake of a key defection and overwhelming evidence to the contrary, Iraq finally admitted to it. In its effort to conceal that program, Iraq had issued false statements, forged documents, and destroyed evidence. According to U.N. weapons inspectors, Iraq was producing two to four times the amount of biological agents it declared, and failing to account for more than three metric tons of material that could be used to produce biological weapons.

It was known that, in the aftermath of the Iran-Iraq war, Iraq had emerged with the largest and most advanced chemical-weapons capability in the Middle East. It repeatedly sought to conceal key elements of this program from inspectors; maintained substantial stockpiles of VX, mustard gas, and other chemical agents; and was rebuilding and expanding facilities capable of producing chemical weapons.

In August 1995, Saddam Hussein's son-in-law, Lieutenant General Hussein Kamel, defected to Jordan. Thanks to the information he provided, Iraq was eventually forced to admit to U.N. inspectors that its previously hidden arsenal contained more than one hundred thousand gallons of botulinum toxin, anthrax, gas gangrene, aflatoxin, and ricin, and almost four metric tons of VX nerve gas.

In the 1990s, U.N. weapons inspectors did succeed in supervising the destruction of much of Iraq's arsenal, including forty thousand chemical munitions, nearly a half-million liters of chemical agents, almost two million liters of chemical precursors, and eight different types of delivery systems, including ballistic-missile warheads. But U.N. officials believed Baghdad had stockpiled far greater amounts of chemical and biological agents than it destroyed.

As for nuclear weapons, in 1995, after four years of deception, Iraq finally admitted it had had a crash program in the field prior to the Gulf War. Had it not been for that war, Saddam Hussein would probably have possessed a nuclear weapon no later than 1993. It was known that Iraq still had much of the infrastructure needed to build a nuclear weapon, and that it had made several attempts to buy high-strength aluminum tubes used to enrich uranium for such a weapon. Had Iraq succeeded in acquiring fissile material, it would have been able to build a nuclear weapon in a matter of months.

In 2002, August Hanning, the chief of the German Federal Intelligence Service (BND), said: "It is our estimate that Iraq will have an atomic bomb in three years." Richard Butler, who headed the U.N. team investigating Iraq's weapons of mass destruction, said: "Saddam Hussein is a homicidal dictator who is addicted to weapons of mass destruction."

Inside his own country, Saddam Hussein's record of terror and repression was as abominable as it could possibly be. Kenneth Pollack, the former director of Gulf affairs at the National Security Council and a senior fellow at the Council on Foreign Relations, has written:

> This is a regime that will gouge out the eyes of children to force confessions from their parents and grandparents. This is a regime that will crush all of the bones in the feet of a two-year-old girl to force her mother to divulge her father's whereabouts. This is a regime that will hold a nursing baby at arm's length from its mother and allow the child to starve to death to force the mother to confess. This is a regime that will burn a person's limbs off to force

him to confess or comply. This is a regime that will slowly lower its victims into huge vats of acid, either to break their will or simply as a means of execution. This is a regime that applies electric shocks to the bodies of its victims, particularly their genitals, with great creativity. This is a regime that in 2000 decreed that the crime of criticizing the regime would be punished by cutting out the offender's tongue.

"Torture is not a method of last resort in Iraq," Pollack states, "it is often the method of first resort." Iraqi expatriates have attested to the truth of these words, as have numerous human-rights organizations like Human Rights Watch and Amnesty International.

Many leaders in the world are evil, and a handful possess weapons of mass destruction. But a toxic confluence of factors—malevolence, aggression, a fondness for terrorists, hatred for America and of course Israel, and an insatiable appetite for weapons of mass murder—makes Saddam Hussein unique. In what Churchill once called "the dark, lamentable catalogue of human crime," his monstrous tyranny has been surpassed by only a few other dictators, all of them rightly reviled in the judgment of history.

Saddam Hussein invaded Iran, invaded Kuwait, attempted to assassinate a former U.S. president, George H. W. Bush, murdered thousands of his own people, and refused to abide by international law. If America strikes against Saddam Hussein, it can eliminate a supporter of terrorism, a tyrant, and a builder of weapons of mass destruction. If we wait, the cost of waiting could be incalculable in terms of the suffering we could have avoided by prompt action now, before he becomes a nuclear power. The primary point is that appeasement—trusting a "homicidal dictator" to keep his word—has never worked, no matter who the homicidal dictator is or was. How many lives, including Iraqi lives, would have to be lost to learn the lesson all over again that appeasement does not work?

In his speech in San Francisco, Al Gore did raise one serious question that others have asked as well: Why "do" Iraq now, when the war against al-Qaeda is still unfinished? It is a question that deserves an answer.

The answer is that September 11, 2001, changed the way we look at American security. As *Washington Post* columnist Jim Hoagland put it, "Knowledge is important not when it first becomes available but when an audience becomes available to absorb and act on the knowledge. Two American presidents in the past decade sank into denial rather than deal with Iraq decisively. September 11, 2001, removed that psychological luxury for George W. Bush and for the American public."

Indeed. That is why, on September 20, 2001, President Bush stated: "From this day forward, any nation that continues to harbor or support terrorism will be regarded by the United States as a hostile regime." And why, in his 2002 State of the Union Address, he declared that states like Iraq, Iran, and North Korea "constitute an axis of evil, arming to threaten the peace of the world. By seeking weapons of mass destruction, these regimes pose a grave and growing danger.

They could provide these arms to terrorists, giving them the means to match their hatred. They could attack our allies or attempt to blackmail the United States. In any of these cases, the price of indifference would be catastrophic."

It is easy to call for peace, and it is easy to say that the answer to an immediate threat to our security is not military action but "containment." But containment was what was long being practiced by the United Nations and by successive American administrations, and if containment had worked, we would not have ended up where we did in November 2002. U.N. resolutions would not have been flouted, inspectors would not have been barred from Iraq for four years, Saddam Hussein would not have been amassing more and more weapons of mass destruction, and he would not be within one to three years from having a nuclear weapon, as informed sources guess—an estimate that, at best, has a "margin of error."

Where was the peace movement while all this was going on, and was known to be going on? How could it sit idly by as a dictator acquired the means of mass destruction, the means to hold the whole world at bay? But sit by it did, waiting to unleash its protest only when the United States appeared resolved at long last to face down the menace and eliminate it. The peace movement has a long record of protesting the nuclear policies of this country. The actor Martin Sheen boasts multiple arrests for such protests. And yet this same movement would rather have America sit idly by than employ conventional force to prevent Saddam Hussein from acquiring a nuclear capability that he himself has sworn to use. The only conclusion one can reach is that the peace movement in this country favors the possession of nuclear weapons by dictators with openly aggressive intentions while opposing their possession by democracies bent on preserving and extending the peace. And it calls itself a *peace* movement?

There is a reason President Ronald Reagan ended the policy of containment and ushered in a policy of rolling back Soviet domination throughout the world. This policy of rollback included more than the invasion of Grenada. It included the funding and training of counterrevolutionary forces in Soviet-sponsored satellites, and the deployment of intermediate-range missiles in Western Europe. American action put an end to the Soviet nuclear threat, and so too the terror under which hundreds of millions had lived during the dark decades of the cold war. The blessings that followed, not just for the peoples of Eastern Europe but for the West as well, are with us still today. One need only ask Nicaraguans or East Germans if they are better off than they were twenty-two years ago under the heels of Communist dictators.

A decision not to act against Saddam Hussein would go down as one of the most dishonorable acts of appeasement and nonfeasance in history. When a dictator is appeased, untold lives are lost, uncountable masses of people are immiserated. But when a dictator is defeated and deposed, the future lies open once again to the energies and the interests of ordinary people, and to the irrepressible forces of freedom. In the case of Iraq, as in the case of the Soviet Union, the blessings that would ensue would redound not just to the people of that country, and not just to the people of the Middle East, but to the whole world.

In Iraq, the responsibility to act is ours because the ability is ours. The responsibility is ours because oppressed people look to us for their deliverance. Abraham Lincoln called our Declaration of Independence "a rebuke and a stumbling block to the very harbingers of reappearing tyranny and oppression." The same could be said not just of our founding document but of our nation as a whole. It is a high honor and responsibility to be entrusted with the might we have earned and deployed on behalf of others and ourselves. Borrowing again from Lincoln, without our might and the willingness to use it, we meanly lose our country and our principles, betraying those who struggle for the ideals and freedoms we ourselves enjoy. With our might and the willingness to use it, we give freedom to the enslaved and defend "the last best hope of earth."

—New Year's Day, 2003